HOME ELECTRICAL WIRING *Made Easy*

Common Projects and Repairs

No. 3072
$24.95

HOME ELECTRICAL WIRING *Made Easy*

Common Projects and Repairs

Robert Wood

TAB BOOKS Inc.

Blue Ridge Summit, PA

FIRST EDITION
FIRST PRINTING

Copyright © 1988 by TAB BOOKS Inc.
Printed in the United States of America

Library of Congress Cataloging in Publication Data

Wood, Robert W., 1933-
    Home electrical wiring made easy.

    Includes index.
    1. Electric wiring, Interior.   I. Title.
TK3285.W66 1988   621.319′′24   88-24781
ISBN 0-8306-0372-7
ISBN 0-8306-9372-6 (pbk.)

TAB BOOKS Inc. offers software for
sale. For information and a catalog,
please contact TAB Software Department,
Blue Ridge Summit, PA 17294-0850.

Questions regarding the content of this book
should be addressed to:

    Reader Inquiry Branch
    TAB BOOKS Inc.
    Blue Ridge Summit, PA 17294-0214

Notices

National Electrical Code is a registered trademark of the National Fire Protection Association.

Wiggy and Qwik-Gard are registered trademarks of the Square D Company.

Signal-Block is a registered trademark of Stanley Automatic Openers.

Light Maker, Signal-Block, and U-Install are registered trademarks of Stanley Automatic Openers.

Disclaimer

It is the intent of the author that the information presented is accurate and reliable; however, because of the nature of electricity, neither the author or TAB BOOKS Inc. is liable with respect to the use of the information contained herein.

Contents

Acknowledgments

It is with sincere appreciation that I thank the following people and businesses for their contributions in the writing of this book:

Rodney B. Beus, Fairmont Electric Service, Boise, Idaho
Square D Company, Lexington, Kentucky
Thomas Industries, Inc., Louisville, Kentucky
Salt River Project, Phoenix, Arizona
Paragon Electric Company, Two Rivers, Wisconsin
Stanley Automatic Openers, Detroit, Michigan

HOME ELECTRICAL WIRING
Made Easy

Common Projects and Repairs

Introduction

THE RESIDENTIAL USES OF ELECTRICITY ARE INCREASING AND NORMALLY provide many years of satisfactory service. Often our needs change or we find areas in our electrical service that need to be upgraded or expanded. The field of electricity is varied and complex, and most electricians spend about four years as an apprentice learning their craft. Good electricians are in the upper part of the pay scale because of their training and experience. They work fast and efficiently. As a homeowner, however, you can take your time and refer to reference material. With a good basic understanding and good work habits, you will be able to make electrical repairs, improve your present wiring, or expand an existing service. This book was written to provide the information to perform these jobs.

You should plan your projects well and coordinate them with the local electrical inspector. Obtain any necessary permits to satisfy building codes. Most people are pleasantly surprised just how simple and easy electrical wiring is. It is important to use the proper materials—this makes the job easier. Anything else just complicates matters.

Although electricity is a science, technical terminology in this book has been kept to a minimum, and illustrations are liberally used. This should enable anyone who is just a little handy with tools to enjoy the satisfaction and save a considerable amount of money by doing his own electrical work. For some, money alone might not be a factor. The immense satisfaction of a job well done is motivation enough.

1

Electricity & Safety

An AFTERNOON BUILDUP OF CLOUDS SUGGESTED SOME RELIEF FROM the summer's heat, but now they continue to grow larger, their bases darkening with moisture. Thunder follows the brief flashes of lightning. In one area of the city, lights flicker then suddenly go out. Television sets go dead. Even time seems to stand still as electric clocks cease their counting. Cash registers in stores refuse to operate, traffic lights are out, and even gas stations are unable to pump fuel. The power company quickly dispatches crews to restore service, while temperatures in refrigerators and freezers begin to inch upward.

Though rare, almost everyone has experienced this situation at some time or other. It strongly emphasizes and reminds us of just how much we depend on electrical energy.

This little-understood phenomenon is the heart of our economy, providing a variety of jobs and electrical servants in our homes, and in general making possible a very comfortable life-style. We need to have a better understanding of this strange energy that provides us with so much power so efficiently.

SOURCES OF ELECTRICAL ENERGY

Our everyday knowledge of electrical energy is based more on what it does, than on what it is. The basic form of electrical energy is of little use. For it to be beneficial, it must be changed into some other type of energy. An electric iron is able to operate when electrical energy is converted into heat. If you apply a current to an electric motor, the electrical energy is changed to mechanical energy. Electrical energy is changed into light because of the resistance of the filament in a light bulb to electrical current. When an automobile battery is being charged, electrical energy is changing to chemical energy.

Electricity is much easier to use than other types of energy. It moves great distances almost instantly. It is a clean, safe form of energy, and as convenient as the nearest light switch.

The source of electrical energy is in the atom (Fig. 1-1). All matter is made up of atoms, and atoms are made up of particles called *protons, neutrons,* and *electrons.* In this book, our interest is in the electrons. Electrons are usually confined to a single atom; however, some can and do move from one atom to another, and they are called *free electrons.* Copper, steel, and aluminum atoms have many free electrons. Consequently, these metals are good conductors of electricity, with copper being the best of the three (Fig. 1-2). Atoms found in materials such as rubber, plastic, paper, and wood have little or no free electrons. These materials will not conduct electricity efficiently, so are classified as *insulators.*

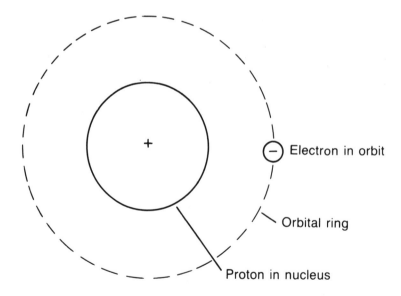

Fig. 1-1. One electron in orbit around one proton in an atom of hydrogen.

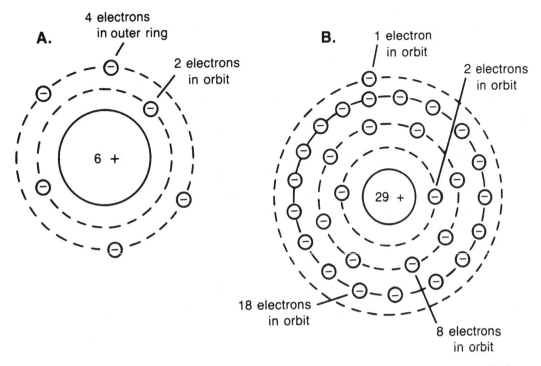

A.

4 electrons
in outer ring

2 electrons
in orbit

6 +

B.

1 electron
in orbit

2 electrons
in orbit

29 +

18 electrons
in orbit

8 electrons
in orbit

Fig. 1-2. Atomic structure: (A) Carbon atom with 6 protons in the nucleus and 6 orbiting electrons. (B) Copper atom with 29 protons and 29 orbiting electrons.

Electricity is everywhere. The most spectacular example is found in the atmosphere as lightning, but you also can feel it in your body as static electricity when you walk across some types of new carpet and then touch a metal door knob. Electricity is colorless, and odorless; it has no size or weight. Electricity, it turns out, is simply a silent stream of electrons in motion. This motion occurs when the balancing force between the protons and electrons is upset by another force. While the electrons in the atoms are moving, they can transmit an electrical charge through solids such as metals and thereby produce an electrical current.

There are two types of electrical current, the simplest being direct current, or *dc* (Fig. 1-3A), which is generally provided by batteries. In direct current, the current flows only in one direction. If on the other hand, current flows first in one direction and then the other, the current is said to be *alternating* (Fig. 1-3B).

Alternating current (ac) is the type of current we use in our homes and is generated by the utility companies. Current flowing in the homes and factories throughout the United States reverses

3

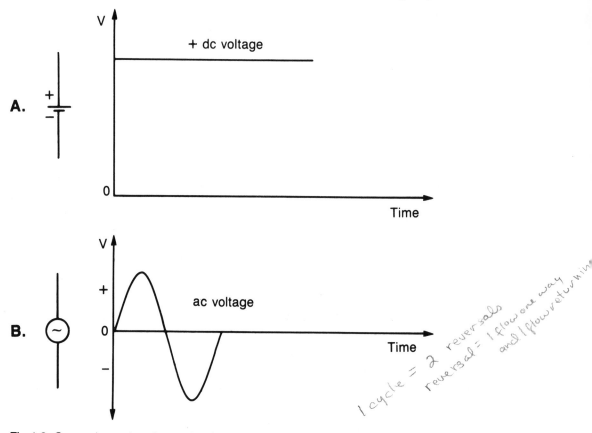

Fig 1-3. Comparing a dc voltage with an ac voltage. (A) a dc waveform showing a steady dc voltage with one polarity. (B) An ac waveform, or sinc-wave, showing an alternating voltage reversing polarity.

1 cycle = 2 reversals
reversal = 1 flow one way and 1 flow returning

itself 120 times a second. It takes two reversals to create or make up a cycle (Fig. 1-4). All homes in the United States operate on the same frequency: alternating current of 60 cycles per second. This frequency is called the *hertz*, and refers to the number of cycles per second.

THE GENERATOR EFFECT—
HOW WE MAKE ELECTRICITY

In 1820, Hans Christian Oersted of Denmark discovered that a strange thing happens when current flows through a wire. A magnetic field is instantly built up around this wire (Fig. 1-5). Normally this field is very small, but if the wire is wound into several coils, the magnetic field of each coil tends to add to the magnetic field of the next coil, and a strong magnetic field is quickly created.

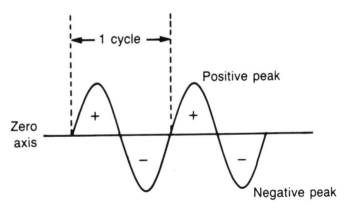

Fig. 1-4. One cycle of an ac waveform.

High-power transmission lines have a magnetic field, which you might have detected on your car radio when passing underneath them. This annoying static indicates a strong magnetic field. The phenomenon provides the principle underlying the operation of solenoids and electromagnets found in many familiar devices, from doorbells and automobile starters to telephones and televisions.

Almost all the electricity we use is produced by generators. Today's generators operate on a principle discovered in 1831 by the English physicist Michael Faraday. After studying the effects of electromagnetism, he believed that <u>if electricity could produce magnetism, then magnetism could produce electricity</u>. He found that a rapidly moving magnet near a copper wire would induce an

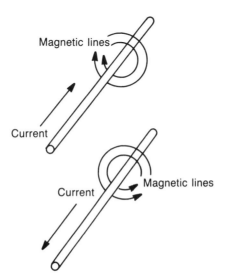

Fig. 1-5. The direction of the magnetic lines depends on which way the current flows.

electrical current in the wire. He further found that by moving the wire rapidly inside a magnetic field, a current also will be induced in the wire.

You can duplicate this strange phenomenon by rotating a loop of wire between the poles of a magnet (Fig. 1-6). When this wire loop, or conductor, passes through the magnetic lines of force between the north and south poles of the magnet, a small current can be detected in the wire. If you look at the ends of a U-shaped magnet, you can visualize the invisible force around each end of the magnet or pole (Fig. 1-7). You can imagine this magnetic field as containing lines of force emanating from the north pole and going

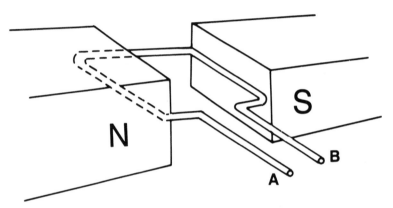

Fig. 1-6. A wire loop forming a conductor between the two magnetic poles.

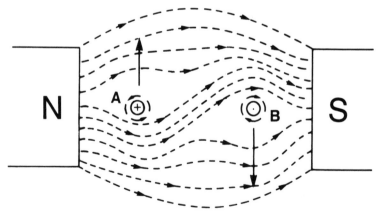

Fig. 1-7. The magnetic field developed around the wire moving through the magnetic field of the magnet, with current entering at (A) and coming out at (B).

back into the magnet at its south pole. The more powerful the magnet, the greater the number of these lines of force.

As the wire loop rotates within the magnetic field between the north and south poles of the magnet, the sides of the loop will cut the magnetic lines of force. This cutting action of the wire conductor through the magnetic lines of force is the phenomenon that induces, or generates, electricity in the wire loop. This process is called *electromagnetic induction*. When the loop begins to rotate, one side passes up through the lines of force, while the other side of the loop moves down (Fig. 1-8). The current in this first half-cycle will flow in one direction. When the loop arrives at the halfway position and neither side is going up or down, none of the lines of force are being cut, and no electricity is being generated.

As the loop continues on into the second half of the cycle, the part of the loop that was formerly moving upward starts moving downward through the lines of force, and the side that was moving downward starts moving upward. The current in the loop begins to flow in the opposite direction of the current induced in the first half of the cycle, when the loop again arrives at its vertical position, again none of the lines of force are being cut, and no electricity is being generated.

During each revolution of the loop, the current flows first in one direction, then reverses and flows in the opposite direction. Twice during this revolution there is no current flowing. This is the basic principle that produces the alternating current commonly found in our homes. The output voltage of this simple generator can be increased by any or all of the following methods: by using a more powerful magnet, which increases the number of lines of force, by adding more loops of wire that cut through the magnetic field, or by rotating these loops faster.

POWER DISTRIBUTION—
GETTING ELECTRICITY TO OUR HOMES

The electricity we use in our homes is supplied to us by the electrical utility companies which convert coal, oil, running water, or atomic energy into electrical energy. These companies operated huge generating stations that send electricity, sometimes great distances, to our cities and homes (Fig. 1-9).

The force that forces electrons to move.

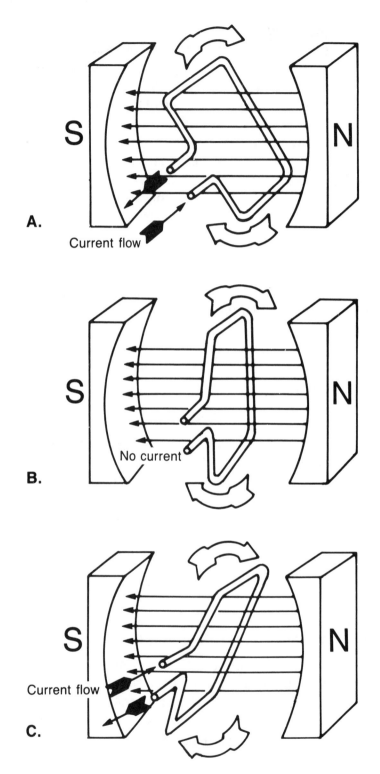

A.

Current flow

B.

No current

C.

Current flow

Fig. 1-8. Wire loop rotating inside a magnetic field. (A) As the loop begins to rotate, current flows in one direction. (B) When the loop is outside the magnetic field, no current is flowing. (C) Current flowing in the opposite direction as loop continues rotation.

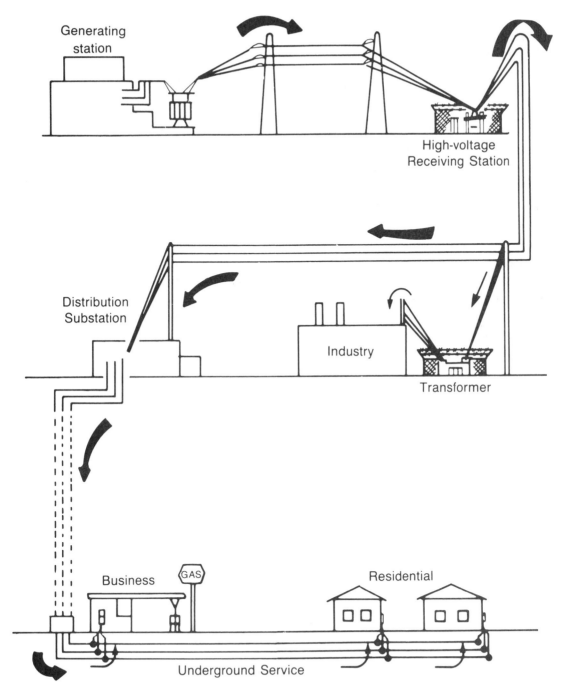

Fig. 1-9. Power distribution. Electricity leaves the generator at 22,000-26,000 volts, passes through transformers, and is stepped up as high as 760,000 volts. Then it enters the transmission system to the receiving stations. From there it travels to substations and may be stepped down to about 12,000 volts for further distribution to consumers.

9

In order to move electricity efficiently over long distances, it must be converted to a very high voltage with a low amperage. The higher the amperage, the larger the conductors must be to transmit electricity. This conversion is accomplished through the use of transformers. After the electricity is generated (Fig. 1-10), it is fed into a *step-up transformer*, which raises the voltage, and as a consequence, lowers the current. Next the electricity flows to the high-voltage transmission lines, where it travels the necessary distance to a substation (Figs. 1-11 and 1-12). At the substation, it is fed into a *step-down transformer*, which lowers the voltage to an acceptable level for the consumers in that area (Fig. 1-13). The final consumers could be factories or industries that use a high voltage of 480 volts or more, or a residential area where another step-down transformer provides the 220 or 110 volts used in our homes.

In some neighborhoods, the service is provided by overhead power lines. In these cases, the canister-shaped devices found on the power poles are the step-down transformers (Fig. 1-14). Underground services used pad-mounted transformers (Fig. 1-15). These line transformers lower the voltage to a manageable level for our homes.

Most people don't realize that when we turn on our air conditioning system or a factory turns on all its lights and starts up its equipment, this increase in electrical consumption is instantly seen at the generating plant, and their generators must respond with a larger supply of electricity. The energy of water can be stored in a reservior or behind a dam. Electrical energy, however, cannot be stored in large amounts. Therefore, power companies must constantly maintain a small army of personnel and equipment to generate electricity the instant the demands are made. Generating plants usually have some form of generating ability on standby. As an example, a generator may be turning but not producing power. Standby generating ability provides the insurance that if needed, extra service can be provided.

When we plug in an electrical appliance and turn it on, the electrons flow from the generating station through the wires to the appliance and return back to the generating station (Fig. 1-16). This movement of electrons is called *current flow*. The amount of current, or the number of electrons that pass a given point, is measured in units called *amperes*, normally shortened to amps. The force, or pressure, that moves these electrons is an electromotive force

Fig. 1-10. Santan generating station located southeast of Chandler, Arizona. Four units produce 75,000 kw apiece.
(COURTESY OF SALT RIVER PROJECT)

Fig. 1-11. Kyrene substation. (COURTESY OF SALT RIVER PROJECT)

Fig. 1-12. A substation is used to step down the voltage for further distribution. (COURTESY OF SALT RIVER PROJECT)

Fig. 1-13. Anderson substation. (COURTESY OF SALT RIVER PROJECT)

Fig. 1-14. Pole-mounted transformer provides overhead service.

Fig 1-15. Pad-mounted transformer providing underground service.

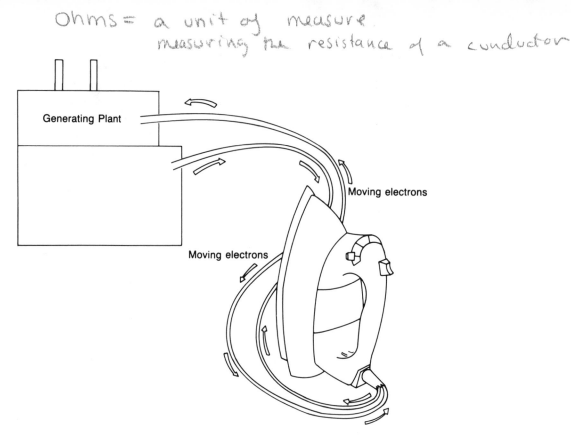

Fig. 1-16. Electrons flow from the generating plant through the appliance and return to the plant.

that is measured in *volts*. The work performed by the voltage and current is measured in *watts*. The most important condition that affects the flow of electrical currents is resistance.

Electricity is selective about the materials through which it flows. As with any current, it will flow in the path that offers the least resistance. This electrical resistance is a phenomenon similar to friction, where the larger the resistance, the smaller the current flow in the conductors. This resistance also has a tendency to generate heat. The resistance of a conductor is measured in units called *ohms*.

You can calculate all these values by a simple formula given by Ohm's Law, which states that the voltage divided by the current equals the resistance, and voltage multiplied by current equals power in watts (Fig. 1-17). To use the formula in the circle illustrated, simply cover the unit you want to find with your finger and perform the calculations of the remaining two.

PUTTING ELECTRICITY TO WORK

For electrical energy to be used, there must be a continuous path or circuit for the current to flow. It often helps to think of electrical circuits as plumbing circuits and water as the electron flow. When the current travels through this circuit,it provides energy that can generate heat, create light,or make motors run.

The simplest electrical circuit consists of a power source and a load (Fig. 1-18). The actual source of power is the light company's generating plant, but for our purposes we'll consider the service entrance to your home as our source of power. The load or circuit can be broken down further into three parts: a conductor or wires, some switching arrangement or power control, and the load of the devices or appliances themselves (Fig. 1-19). These appliances could be a variety of devices from lamps, radios, and televisions, to telephones and motors. The switch is simply a device that controls the flow of the current. In the example of water in plumbing, this could be the faucet to a sprinkler system in the lawn.

Because of the large volume of matter making up our Earth, the Earth is electrically neutral. Consequently, current will flow when voltage is applied to a wire or conductor connected to the Earth or ground. If this conductor happens to be the human body, severe shock will occur (Fig. 1-20). However, by using common sense and following a few simple rules, you can handle electricity quite safely and easily.

ie: earth is a good conductor

GROUNDS AND SHOCK HAZARDS

Electricity is something everyone uses quite freely with just the flip of a switch or the push of a button, but it is also something that should be treated with respect. Occasionally the news media reports on fires or injuries resulting from electrical causes. In rural areas, barns have burned to the ground because of electrical storms. Houses and mobile homes are sometimes completely destroyed because of faulty wiring. People are shocked, and in severe cases, have died from household electrical accidents.

There is little you can do about electrical storms, but you should never be careless with electricity in your home. Fires in the home are usually caused by faulty wiring, which might be caused by nothing more than a poorly connected plug to a cord. Any poor connection could lead to overheating and eventually cause a fire.

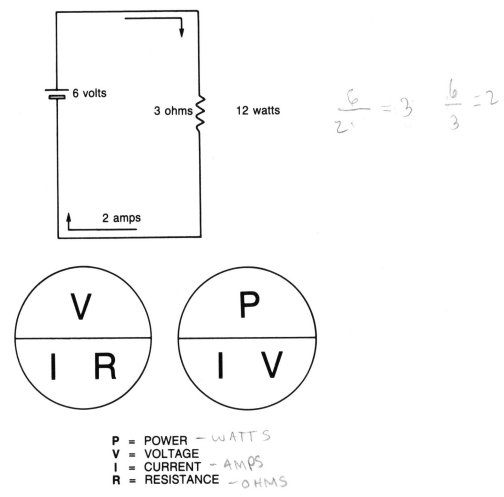

$$\frac{6}{2} = 3 \qquad \frac{6}{3} = 2$$

P = POWER — WATTS
V = VOLTAGE
I = CURRENT — AMPS
R = RESISTANCE — OHMS

Fig. 1-17. Ohm's Law. In this illustration, a 6-volt supply operating a circuit using 2 amps will produce 12 watts of power.

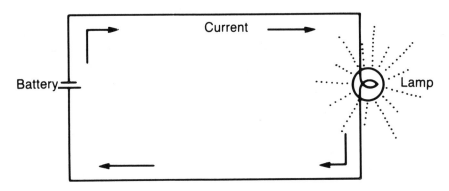

Fig. 1-18. A simple electrical circuit.

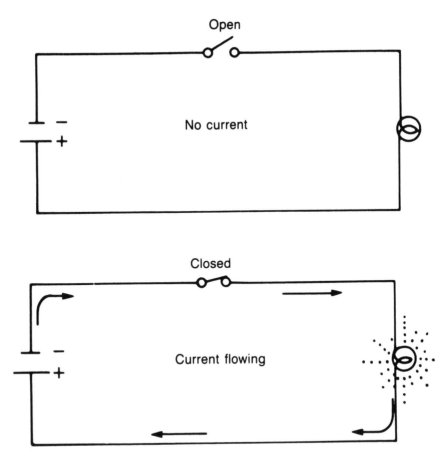

Fig. 1-19. A circuit must have a complete path for current to flow. (A) Switch open. (B) Switch closed.

Wires also can overheat when too many appliances are plugged into an extension cord that isn't large enough to carry the electricity the appliances demand. As the heat builds up in the cord, the cord insulation melts or becomes brittle. As the insulation deteriorates the wires become exposed. It is only a matter of time until the bare wires touch each other, causing a short circuit, and sending sparks flying. Therefore, any appliance cord or extension cord in use that becomes warm to the touch should be considered a potential fire hazard.

Getting an electrical shock is akin to getting bitten by a snake. It happens fast and is usually a surprise. Keep in mind, however, that conditions must be exactly right for you to get an electrical shock. Just touching a wire won't necessarily do it. Remember that

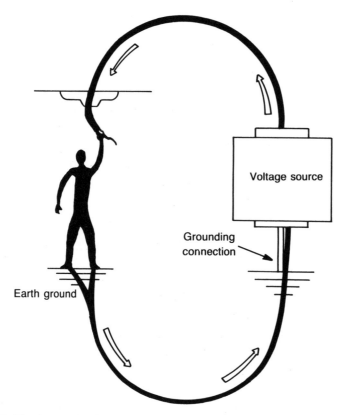

Fig. 1-20. The human body can complete a circuit, allowing current to flow.

current must flow in a continuous closed path from its source through some device or load and then back to the source. If for some reason, you happen to become the link in an electrically live circuit, you will receive a shock. You must make the connection. You must be touching the live wire and at the same time come in contact with a grounded object or another live wire. Your body must form the link or make the connection to complete the circuit. All of this means that electricity does not need to flow in wires to make the return trip to its source. Electricity can return to the source through any conductor, including the human body, that comes in contact with the earth directly or comes in contact with another conductor that in turn touches the earth.

At first, this situation might seem remote, but remember that if you are taking a bath or swimming in a pool, touching any metal water pipes or faucets, or standing on the ground or on a damp concrete garage floor, you become a conductor to ground. Now all you have to do is touch a hot wire or come in contact with a live circuit, and a shock will most certainly occur. To avoid this situation, it is only necessary to observe one of the most important rules for any do-it-yourself electrician. "Never,work on any electrically live circuit, fixture or appliance." Your safety and your life might depend on how well you obey this one rule.

Before reaching for a screwdriver or wire strippers, simply disconnect, or kill, the circuit you are working on. Do so at its source—at the fuses or circuit breakers in the service entrance panel. If your service entrance panel uses fuses, simply remove the appropriate fuse to disconnect the circuit from the incoming service. If your service panel uses circuit breakers, moving the breaker to the OFF position disconnects the circuit. In order to ensure that you have killed the right circuit, turn on an overhead light or plug a lamp into a wall outlet before killing the circuit, keeping in mind that wall outlets are normally not on the same circuit as the overhead lights. The overhead light or the lamp will go out if you kill the correct circuit, but if you still, have any doubt about which fuse or breaker kills which circuit, you can kill all the power by shutting off the main breaker in the entrance panel. With the main breaker off, take a look at the meter. The thin wheel or disk lying horizontally below the dials should not be turning.

Industrial electricians have padlocks for lock-outs and use red tags to identify circuits that are being worked on. In this case, a note taped to the panel explaining why the power is off could prevent a possible disaster. Someone could come along and reset the breaker or replace the fuse. For added insurance, tape the circuit breaker to the OFF position: if your house uses fuses, take the fuse with you. Now that you have killed the circuit and are confident that the power is off, make one final check to determine that the circuit is actually dead. Use a meter or test lamp. When you are satisfied that the power is completely off, you can begin your wiring project.

However, you need to keep in mind a few additional safety precautions.

Don't forget that water is a good conductor. Never do any electrical work while you are wet, or standing in a damp location. If the ground or floor is wet, put down some dry boards as a temporary work platform.

Spend some time thoroughly familiarizing yourself on how your particular home is wired before you modify or add any circuits to the electrical system.

Always be on guard against sloppy wiring practices. the person who was there before you might not have observed proper wiring techniques. Labels in the panel might be misleading.

Keep in mind that even though the main circuit breaker might be tripped in the service entrance panel, the wires coming in from the meter feeding the panel are still hot and must be respected.

If you do encounter a situation where someone has been shocked and they are still part of the electrical circuit, do not touch them with your bare hands. First, try to kill the power: If this is not possible, use some sort of insulating device such as a coat or a broom to remove the victim from the circuit. Then keep the victim warm and give artificial respiration, if necessary, until help arrives.

By far the greatest safety device you have is your own mind. Stop and think, work slowly, and if you are in a hurry, don't do it. With proper planning and good work habits, most do-it-yourselfers are pleasantly surprised just how easy electricity is to work with, and soon are thinking of new and original ways of improving their home-wiring system.

2

Staying
Safe and Legal

NEARLY 100 YEARS AGO, A GROUP OF INSURANCE, ELECTRICAL, ARCHI-
tectural, and associated interests developed the original code that
became known as the National Electrical Code. It has been
sponsored by the National Fire Protection Association (NFPA) since
1911. The NFPA continues to publish revised and updated editions
of the code every three years to stay abreast of new developments
in the field of electricity. A technical committee makes rec-
ommendations based on the primary objective of safety and fire
prevention.

The code itself is not a law, only an advisory. However, it is
usually adopted into law by most cities and counties. These
governing bodies might further adopt local ordinances that pertain
to their particular area.

Professional electricians must pass an examination based on
the rules and regulations of the National Electrical Code, along with
the local code for their area.

Most bookstores have copies of the National Electrical Code.
Few people can quote from it in its entirety, and as a homeowner
you need only be familiar with the regulations that pertain to your

wiring project. Always check with your city or county building inspection department. Some wiring projects, such as hot tubs, swimming pools, etc. might require the services of a licensed electrician.

DOING YOUR OWN WIRING

Some modifications and additions to existing wiring usually can be done by the resident homeowner, but they normally require a permit from the building department and inspections by the city electrical inspector. Permit fees vary from $5.00 to $50, depending on the size of the project. Keep in mind the success of any project depends on how well it is planned.

Using the proper materials is very important. Circuits of 110/120 volts should use wires rated at 600 volts. Normally the wiring in houses is considered to be in a *dry location*. These cables usually consist of three conductors: two wires and a ground wire. You may locate some circuits in damp locations, such as under canopies, patio covers, and anyplace where moisture can come in contact with the wiring (such as storage sheds, barns, and other outbuildings). Damp resistance wiring must be used in these locations.

A *wet location* is any location where water might come in direct contact with the wiring, for example, underground installations and wiring exposed to the weather. Always use the proper cables rated for the conditions at their locations. Your safety and peace of mind is well worth the little extra cost, and the work would not pass inspection otherwise.

WIRE AND CABLE SIZES

The size of the cable is determined by the amount of current it is to carry. American Wire Guage (AWG) wire sizes come in numbered sizes; where the smaller the number, the larger the wire (see Table 2-1 and Fig. 2-1). The receptacle might be fed by #12 wire, while a clothes dryer will require a larger size, such as #8 or #6. Cables are further identified by the number of wires in the cable, such as ''12/2 with ground,'' which means the cable is made up of two #12 conductors and a ground wire (see Figs. 2-2 and 2-3).

The code states that the *overcurrent protection*, or breaker size, for copper cables must not be greater than 15 amps for #14, 20

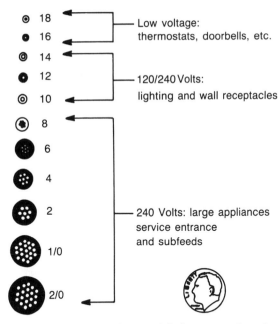

Fig. 2-1. Cross sections of copper wires and their comparative size.

Table 2-1. Current-Carrying Capacities of Copper Wire.

AMPACITY OF COPPER WIRE

	In Conduit, Cable, or Buried Directly in the Earth		Single Conductors in Free Air		
Wire Size	Types T, TW	Types RH, RHW, THW	Types T, TW	Types RH, RHW, THW	Weather-proof
14	15*	15*	20*	20*	30
12	20*	20*	25*	25*	40
10	30	30*	40	40*	55
8	40	50	60	70	70
6	55	65	80	95	100
4	70	85	105	125	130
2	95	115	140	170	175
1/0	125	150	195	230	235
2/0	145	2175	225	265	275
3/0	165	200	260	310	320

* Higher ampacities are given in the code book but the above figures are to be used for overcurrent protection.

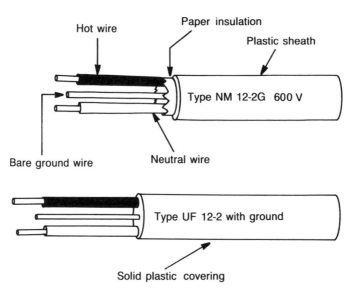

Hot wire
Paper insulation
Plastic sheath

Type NM 12-2G   600 V

Bare ground wire
Neutral wire

Type UF 12-2 with ground

Solid plastic covering

Fig. 2-2. Two types of nonmetallic sheathed cable. Type NM is used only in dry indoor locations, where Type UF may be buried directly in the ground as well as anywhere Type NM is used.

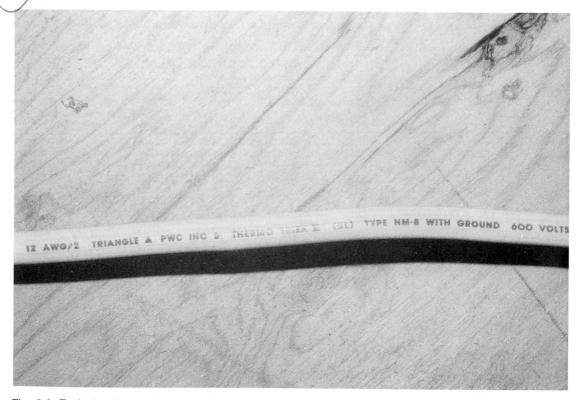

Fig. 2-3. Typical cable used in residential wiring.

24

amps for #12, and 30 amps for #10 wire. Although the code allows #14 cable to be used in some cases, most homes currently being built use #12 as the smallest cable.

Aluminum and copper-clad aluminum wires were used when copper was expensive, but currently copper is inexpensive and copper wire is by far the recommended conductor for today's wiring.

THE COLOR CODE

grounded = commoN Neutral white Natural grey

The code states that residential wiring will have a ground conductor, and that the color will be white or natural gray. This wire is usually grounded by the utility company, and is thought of as the common, or *neutral* wire in the electrical system. Another ground connection is made through the residential water pipes and further may be made through an 8-foot metal grounding stake driven into the ground near the service entrance (Figs. 2-4 and 2-5).

Equipment ground is identified by green insulation, or it can be a completely bare wire. This wire is provided for the user's safety.

Think of the black wire as delivering the electricity to the outlet, the white wire as returning it to the generating plant, and the bare ground wire as tripping a breaker if anything goes wrong with the appliance. The neutral wire also will trip a breaker if it is connected to the hot wire without a load.

CABLE AND CIRCUIT PROTECTION

It is necessary to protect cables from damage that could occur from future construction. When you run a cable through holes bored in joists or rafters, you must bore the holes so that the distance from the edge of the hole to the outside edge of the wood is at least 1¼ inches. If the distance is less than 1¼ inches, then you must install a steel plate at least ¹⁄₁₆ inch thick to protect the cable from being penetrated by nails or screws used when Sheetrock or paneling is installed (Fig. 2-6).

You may install cables in notches in wood studs and joists if the notches don't weaken the structure, provided the notched area is covered by a steel plate at least ¹⁄₁₆ inch thick for protection from nails or screws (see Fig. 2-7).

If the cable enters or passes through a metal opening, a factory or field-drilled hole or knock-out, you must securely install a bushing, or a grommet in the opening before you pull the cable through (Fig. 2-8).

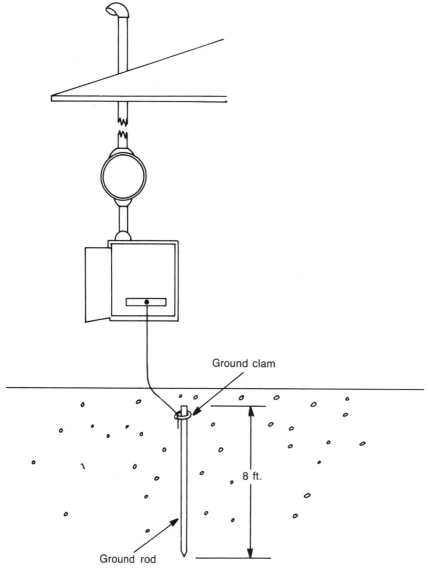

Ground clam

8 ft.

Ground rod

Fig. 2-4. Grounding system using a ground rod.

Buried cables must have minimum cover, depending on the type of installation (Table 2-2). Measure the cover from the finished grade of the surface to the top of the cable or conduit, not to the bottom of the trench. A 24-inch cover does not mean a 24-inch trench.

Fig. 2-5. Grounding rod with clamp.

There are exceptions to Table 2-2. For example, a circuit that controls a low-voltage outdoor lighting system or a sprinkler system of not more than 30 volts and that is installed with the proper underground cable may be installed with minimum of 6 inches of cover.

Another exception is a residential branch circuit to be installed with overcurrent protection of not more than 30 amps and rated 300 volts or less. Then the minimum cover would be 12 inches.

Do not use any backfill containing sharp rocks or materials where they might damage the cables or conduit. You can provide protection by using a granular material, such as sand, or by using suitable running boards or sleeves.

GROUND-FAULT DETECTION

Ground-fault circuit interrupters disconnect or kill an electrical circuit when the current to ground exceeds a very small predetermined amount (Fig. 2-9). This amount is considerably less than the amount of current required to trip a circuit breaker or cause

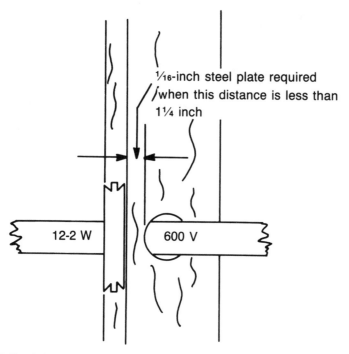

¹⁄₁₆-inch steel plate required
when this distance is less than
1¼ inch

12-2 W

600 V

Fig. 2-6. Steel plate provides protection if the hole is too close to the edge of the stud.

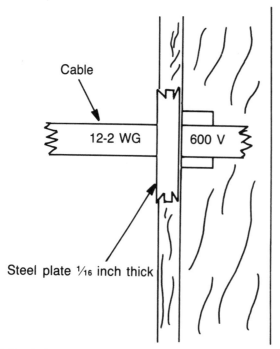

Cable

12-2 WG

600 V

Steel plate ¹⁄₁₆ inch thick

Fig. 2-7. Steel plate is used to protect wires when notches are used.

Table 2-2. Minimum Cable Depths for Underground Wiring.

MINIMUM COVER REQUIREMENTS, 0 TO 600 VOLTS	
WIRING METHOD	**MINIMUM BURIAL IN INCHES**
Direct Buried Cables	24
Rigid Metal Conduit	6
Intermediate Metal Conduit	6
Rigid Nonmetallic Conduit	
Approved for Direct Burial	18

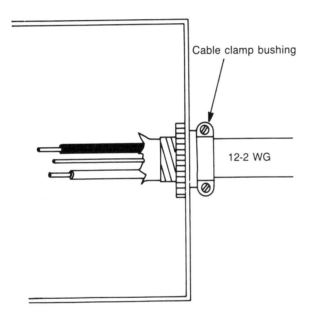

Cable clamp bushing

12-2 WG

Fig. 2-8. Bushings must be used when cables enter metal boxes.

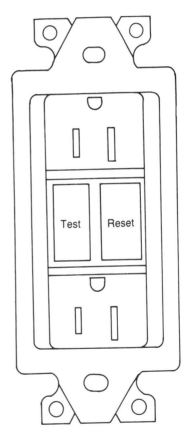

Test   Reset

Fig. 2-9. Ground-fault circuit interrupters provide additional safety for residential users.

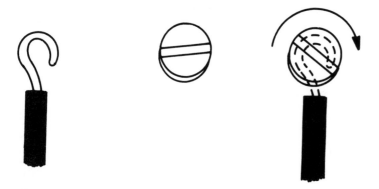

Fig. 2-10. Wire loop should be installed clockwise under screw heads.

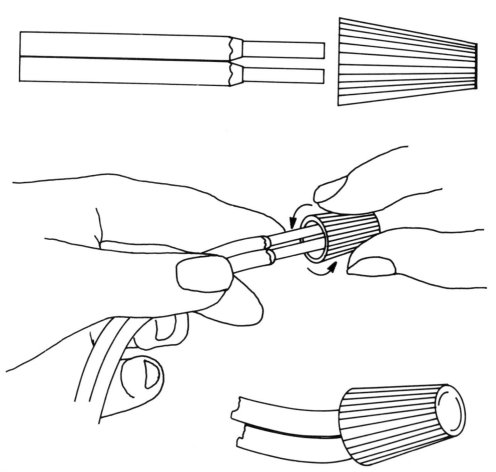

Fig. 2-11. Wire nuts are twisted clockwise when connecting wires. (COURTESY OF SQUARE D COMPANY)

a fuse to blow. These receptacles have a small button on the front that you can use to reset the circuit when the fault has been corrected.

Ground-fault circuit interrupters are now required where the user might come in contact with water or water pipes such as in bathrooms, carports, garages, or outdoor locations within 6½ feet of the ground. Basements are also required to have at least one of these receptacles. Also included in this class are any receptacles above the countertop and within 6 feet of the kitchen sink, except for specific receptacles used for refrigerators and freezers. Note that each individual outlet does not need to be a GFCI receptacle, but each circuit must have GFCI protection. An outlet in the garage on the same circuit as a bathroom with a GFCI receptacle would be satisfactory. Any outside outlet available to the homeowner—for the use of a drill in the garage, to an electric lawnmower or an electric grass trimmer—must be so protected.

Many problems occur in circuits because of loose or poor connections. Most of these problems could have been avoided by proper connections at the beginning (Figs. 2-10 and 2-11). Dissimilar metals can cause heating problems. Avoid copper-to-aluminum connections and make certain that any connection is mechanically solid.

An inexpensive device for the protection of small children is a safety plug (Fig. 2-12). These plastic snap-in plugs are available at most electrical supply houses. When they are snapped in place on unused outlets (Fig. 2-13), they offer protection from a child inserting anything into the receptacle and prevent what could be a severe shock or burn.

Don't think of the National Electrical Code as a hindrance to your project. It is a reflection of nearly 100 years of experience and problem solving in the use of the phenomenon we call electricity. Remember, its primary objective is to keep you and your property safe.

Most homeowners take pride in their projects, enough so that their own standards often surpass the codes for their area. If the code calls for a minimum of a #14 wire, you will always feel better if you have used #12. The code minimums might not always meet the requirements of a conscientious homeowner who has his life savings invested in his residence.

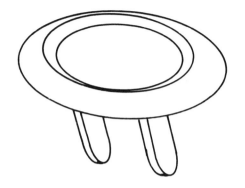

Fig. 2-12. Plastic plugs provide protection on unused receptacles in reach of small children.

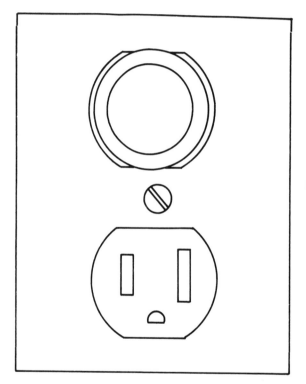

Fig. 2-13. Safety plug installed in receptacle.

Look at the code as a reference point for the safety of your wiring project. If you follow or exceed the code, then you will never have any problems with inspectors. You'll want to perform your wiring project in such a manner that once you have made the last connection and applied the power, you'll never be troubled by that installation again. It will be trouble-free from now on.

There are motors still running today that are well over 40 years old because the installer did it right the first time. You don't have to get paid for the job to do professional-quality work.

white,
grounded

Basic Circuits

UTILITY COMPANIES DELIVER POWER TO OUR HOMES THROUGH OVER-
head wires or an underground service. The standard service
consists of three wires: two hot wires, or conductors, and one neutral
wire (Fig. 3-1). Each hot wire supplies 120 volts with respect to the
one neutral wire. The neutral wire is kept at 0 volts and is considered to be at ground potential.

With this arrangement, one hot wire and the neutral wire will
provide 120 volts for circuits used for such things as lights and wall
outlets. Both hot wires and the neutral provide 240 volts for
appliances, such as ranges and dryers. The hot wires can be any
color except white or green, but normally both are black or black
and red. The neutral wire will be white or might even be bare.

DELIVERING POWER

The utility company connects the incoming wires to a
weatherproof cabinet (Fig. 3-2) that will hold the meter. The two
hot wires and the gray, neutral wire are next fed through the opening
and through the wall into the service entrance panel (Fig. 3-3). The

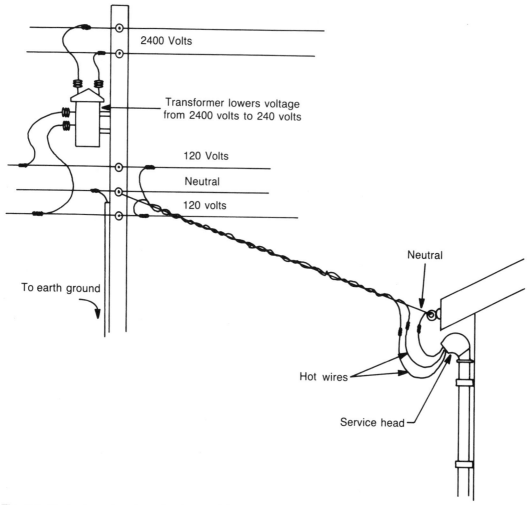

2400 Volts

Transformer lowers voltage
from 2400 volts to 240 volts

120 Volts

Neutral

120 volts

Neutral

To earth ground

Hot wires

Service head

Fig. 3-1. Typical overhead service to a residence.

wires are now connected to their proper terminals (Fig. 3-4) and are ready for the meter to be attached.

As shown in Fig. 3-5, a meter is installed to measure the electricity as it enters the service entrance panel. This meter is calibrated in kilowatt-hours. It measures the amount of energy consumed in kilowatts according to the number of hours used. Most meters have five numbered dials with pointers (Fig. 3-6). The first dial is numbered clockwise, and the rest alternate counterclockwise and clockwise.

To read a meter, write down the numbers from left to right, beginning with the dial on the left. Use the smaller number when

Fig. 3-2. Weatherproof cabinet for the electric meter.

the pointer is between two numbers. If the pointer is directly on a number, look at the next dial to the right. If that pointer is on zero or has passed zero, use the number at the pointer of the first dial. If the pointer of the second dial hasn't reached zero, use the next smaller number on the first dial. In Fig. 3-6, the pointer of the dial on the left is between 1 and 2, so the number you use is 1. On the next dial, the number is between 3 and 4, so you use 3. The next dial's pointer is almost at 3, so a quick check at the pointer of the fourth dial tells you that because it has passed 9 but has not reached

Fig. 3-3. Inside view of cabinet showing wires going to the service entrance panel on the other side of the wall.

Fig. 3-4. Wires connected to the meter socket. Utility company can now bring the service wires up through the conduit, make the connections at the terminals at the top, and attach the meter.

zero. You will use the 2 on the third dial. The number on the fourth dial will be 9 and the final number from the fifth dial will be 6.

To determine the number of kilowatt-hours of a particular period, subtract the reading taken at the beginning of that period from the reading taken at the end.

After the meter, the two hot wires travel into the service entrance panel (Fig. 3-7), where they are connected to the main disconnect or breaker (Fig. 3-8). The neutral wire is connected to the neutral busbar, which is bonded to the cabinet (Fig. 3-9). Fig-

Fig. 3-5. An electric meter installed by the utility company.

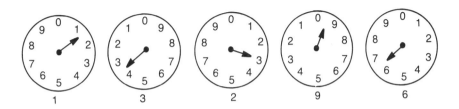

1   3   2   9   6

13296 kilowatt-hours

Fig. 3-6. Numbered dials indicating the amount of electric power used.

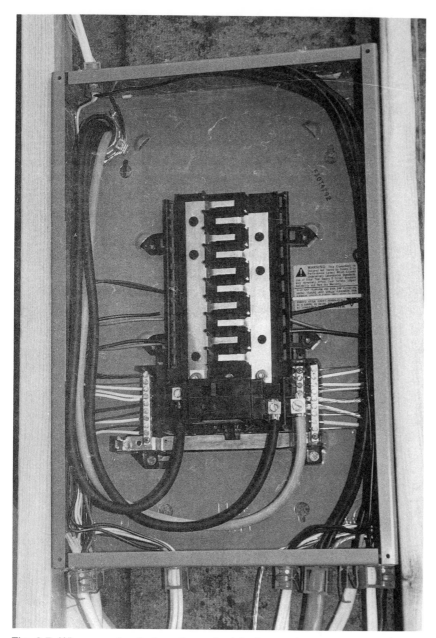

Fig. 3-7. Wires run directly from the meter into the service entrance panel.

ure 3-10 illustrates another important wire connected to the service entrance panel: the ground conductor. This ground wire (Figs. 3-11 and 3-12) connects the neutral busbar to a permanently grounded object, such as a cold water pipe or grounding stake. This connection establishes an electrical path directly to the earth for

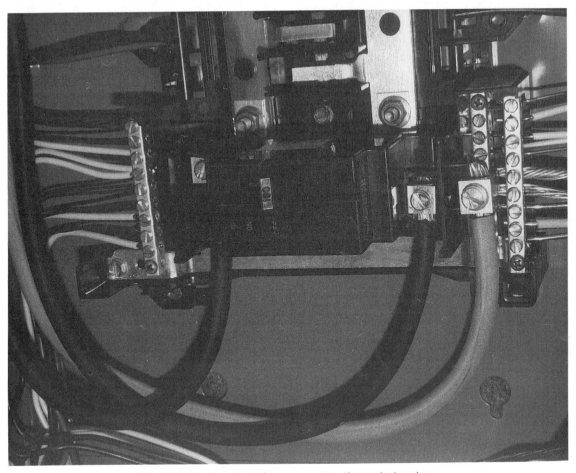

Fig. 3-8. The two hot wires connect to the main breaker.

the complete electrical system.

From the main disconnect, two conductors are connected to two hot busbars. These busbars are able to handle the current allowed by the main disconnect and permit the distribution of the current into smaller currents for each branch circuit. The hot wire of a 120-volt branch circuit is connected to one of the hot busbars through some overcurrent protection device, normally a circuit breaker or fuse (Fig. 3-13). The gray or white neutral and the bare ground wire are connected to the neutral busbar in the cabinet. A a 240-volt circuit breaker to the hot busbar. All neutral and bare ground wires originate at the neutral busbar. There they are in direct electrical contact with the earth or ground. A neutral wire must never be interrupted by a fuse or circuit breaker.

Fig. 3-9. The neutral wire is connected to the neutral busbar.

BRANCH CIRCUITS

A simple light circuit is illustrated in Fig. 3-14. It is only a partial circuit because the switch and equipment ground wire are not

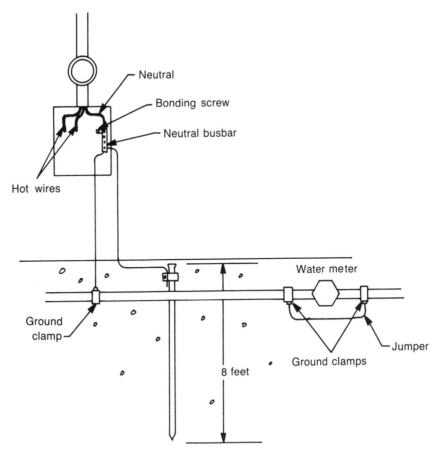

Fig. 3-10. Grounding methods for service entrance panel.

shown. Figure 3-15 shows the same circuit with a switch. Switches should only be installed in the hot wire, not in the neutral one. A look at the illustration will show that the switch will disconnect the device completely from the hot busbar, eliminating the danger of a shock or short circuit if the switch is open. A switch in the neutral wire (Fig. 3-16) also would turn the device off, but it would still be connected to the hot busbar and remain a potential shock hazard.

The equipment ground wire, usually bare, also helps prevent shocks. Normally it does nothing, like a lifeboat; however when a device malfunctions, it instantly goes to work. If any part of the hot wire somehow came in contact with a metal fixture or housing, the housing would be electrically live. A human body could provide the electrical path to ground for current to flow, and someone could receive a severe shock (Fig. 3-17). This situation can occur when

Fig. 3-11. View of the bare ground conductor running from the bottom of the panel, alongside the stud, out through the wall, and down to the ground rod.

Fig. 3-12. Additional grounding is provided when the ground conductor is clamped at a bushing.

a person is operating any metal power tool or appliance. With a ground wire connected from the neutral busbar to the metal fixture or housing, an alternate and less resistant path to ground is provided; in the event of a short circuit, the breaker would trip, turning off the power (Fig. 3-18).

Several branch circuits might be needed to supply power to a kitchen area (Fig. 3-19), and a larger number is required for an entire home; however, by using numbers and electrical symbols, you can make a working drawing or map of an electrical system (Fig. 3-20). In this illustration, the wires are not shown, but a dotted line indicates which switch controls what fixture.

The circuits feeding the receptacles or outlets are connected to the service entrance panel, as shown in Fig. 3-21. Note that the ground is maintained throughout the circuit. This can be very

Fig. 3-13. Circuit breaker to be installed for overcurrent protection.

important in the event of a malfunction in the system or in some device plugged into the system. The overcurrent protection will come into play, and a fuse will blow or circuit breaker will trip and kill the circuit.

CIRCUIT BREAKERS AND FUSES

The purpose of a circuit breaker or fuse is to protect electrical circuits from damage by too much current. By design, a circuit breaker will trip or a fuse will blow if the circuit it is protecting is forced to carry more current than the wiring can safely handle. These excessive currents can come from surges from the utility company or lightning, but more often they are caused by a faulty appliance,

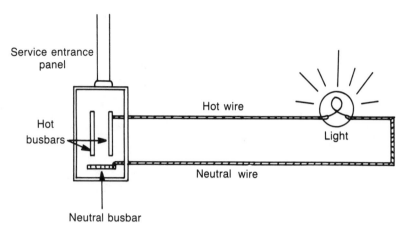

Fig. 3-14. A simple circuit showing the connections in the service entrance panel.

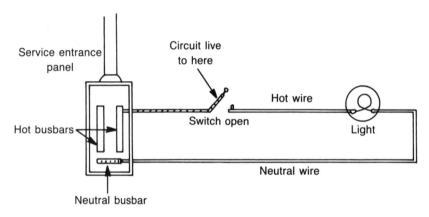

Fig. 3-15. Circuit showing the switch installed in the hot wire.

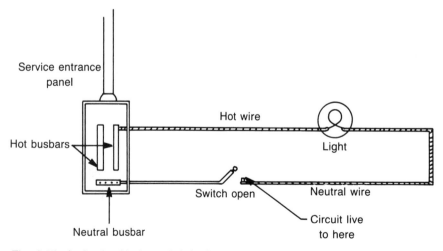

Fig. 3-16. A circuit with the switch in the neutral wire will leave the fixture live.

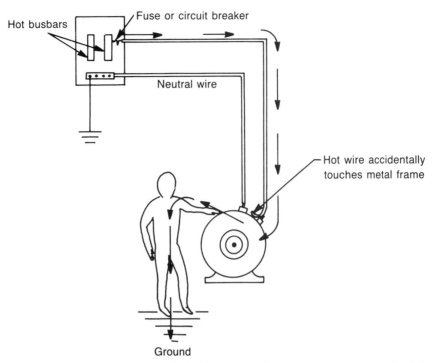

Hot busbars

Fuse or circuit breaker

Neutral wire

Hot wire accidentally
touches metal frame

Ground

Fig. 3-17. Metal appliances installed without an equipment ground present potential shock hazards.

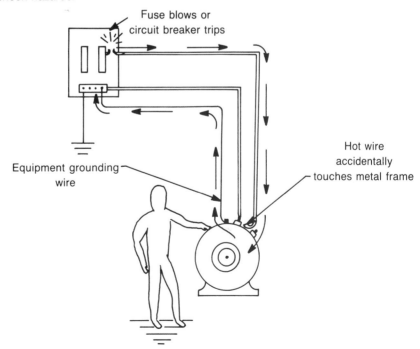

Fuse blows or
circuit breaker trips

Equipment grounding
wire

Hot wire
accidentally
touches metal frame

Fig. 3-18. Equipment ground wires offer a least-resistance path to ground.

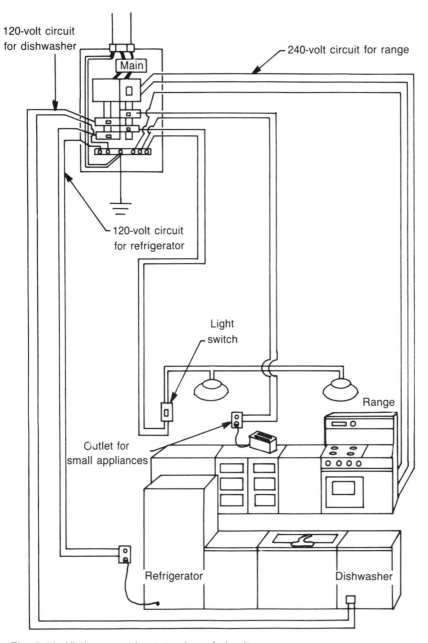

120-volt circuit
for dishwasher

240-volt circuit for range

Main

120-volt circuit
for refrigerator

Light
switch

Range

Outlet for
small appliances

Refrigerator

Dishwasher

Fig. 3-19. Kitchens require a number of circuits.

a bad switch, or a bad receptacle. Too many appliances on one
circuit also will cause a breaker to trip.

To provide proper protection, the fuse or circuit breaker must
be rated for the same current as the wiring it is protecting. For

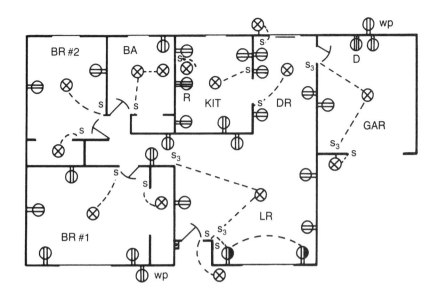

Electrical Symbols

⊗	Light Fixture	⊜	Range Outlet
⊜	Duplex Receptacle	⊜	Dryer Outlet
⊜	Duplex Receptacle, half controlled by switch	▭	Doorbell
S	Single-pole Switch	⊜ wp	Weatherproof Receptacle
S₃	Three-way Switch	- - - - -	Switch Wiring

Fig. 3-20. Electrical symbols used in residential wiring.

example, #12 copper wire can safely handle 20 amps. The fuse or breaker must never be larger than 20 amps otherwise, the wiring becomes the fuse, and if a problem occurs, the wires could overheat enough to cause a fire inside a wall or attic. Never replace a fuse or breaker with one with a higher rating unless you are dead certain the wires are big enough.

A *fuse* is simply a small strip of a metal alloy that has a low melting point. When properly installed, it becomes the weakest link in the electrical circuit. If the current begins to exceed the rating of the fuse, the metal alloy melts and breaks the circuit (Fig. 3-22). The Edison-base fuse no longer can be used in new construction but is permitted as a replacement. An improved version of this type of fuse is the type ''S'' fuse (Fig. 3-23). This fuse has an adapter

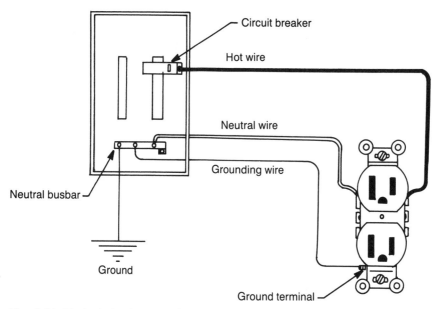

Fig. 3-21. Typical circuit connections in the service entrance panel.

that must be installed. This configuration limits the size of the fuse to the proper rating of the circuit.

Cartridge fuses come in two basic styles: ferrule and knife-blade (Fig. 3-24). The *ferrule* fuse will normally be found protecting the wiring of an individual appliance such as a range. They come in sizes from 10 to 60 amps in the 120/240-volt class. The *knife-blade fuse* sometimes can be found as the main disconnect in fused service-entrance panels. They come in sizes of 70 amps or more and are rated for 240 volts.

For the most part, circuit breakers are replacing fuses (Fig. 3-25). They vary from low-current, pop-out buttons protecting electronic circuits to units resembling light switches in the 20-amp range and the large 200-amp main disconnects found in service-entrance panels. Circuit breakers provide automatic overcurrent protection and are more convenient when disconnecting the power to make repairs or additions to a circuit.

Unlike fuses, which self-destruct when their rated current is exceeded, circuit breakers have a bimetallic strip that bends when exposed to excessive heat (Fig. 3-26). When this strip bends, it releases a spring-loaded trip, which trips and opens the circuit (Fig. 3-27). The breaker then goes to OFF or an intermediate position.

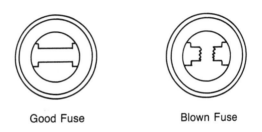

Good Fuse   Blown Fuse

Fig. 3-22. Fuses protect a circuit by allowing a metal strip to melt if the current becomes excessive.

Fig. 3-23. S-type fuses use adapters to restrict the fuse to the proper size for that circuit.

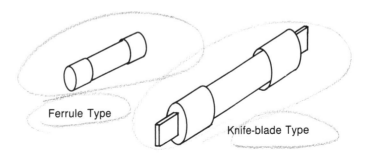

Ferrule Type

Knife-blade Type

Fig. 3-24. Cartridge-type fuses.

After the strip has cooled and the problem is fixed, most breakers can be reset by forcing the handle beyond the OFF position and then moving it to ON (Fig. 3-28). Some breakers only move to the OFF position when tripped, they need only to be returned to ON to be reset.

You may place the service-entrance panel on an outside wall and exposed to the elements. In such cases, however, you must locate them within a rainproof enclosure (Fig. 3-29). You also must

Fig. 3-25. Typical circuit breaker installed in today's homes and factories. (COURTESY OF SQUARE D COMPANY)

make sure the conduit connection is kept rainproof (Fig. 3-30). An indoor service-entrance panel (Fig. 3-31) does not have to be so protected, but both types should be easy to get to in an emergency. Keep obstructions, furniture, storage boxes, etc., away from the front of the panel.

At one time or other, most homeowners will have to deal with a blown fuse or tripped breaker. It becomes expensive and time-consuming to call in an electrician every time the lights go out or the toaster doesn't work. By becoming familiar with the basic residential electrical system, almost anyone can be a little more electrically independent and give his household maintenance budget a break.

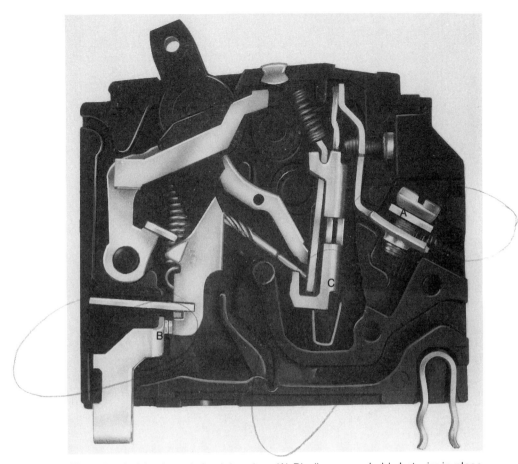

Fig. 3-26. Inside view of circuit breaker. (A) Binding screw holds hot wire in place. (B) Electrical contacts. (C) Thermal-magnetic tripping mechanism. (COURTESY OF SQUARE D COMPANY)

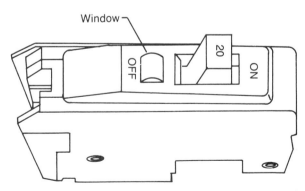

Fig. 3-27. Some circuit breakers have little windows that indicate when they are tripped.

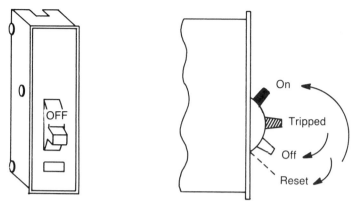

Fig. 3-28. Method of resetting a tripped circuit breaker.

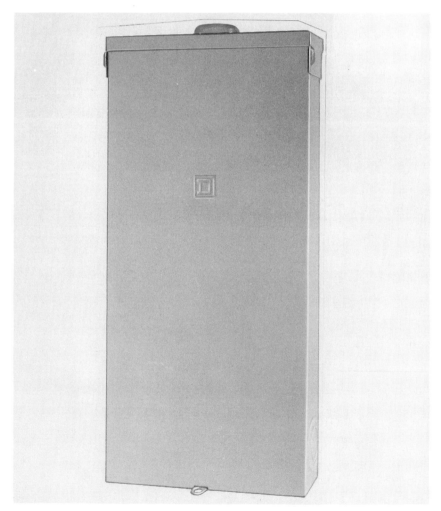

Fig. 3-29. Rainproof enclosure. (COURTESY OF SQUARE D COMPANY)

Fig. 3-30. Bolt-on hubs make rainproof conduit connections. (COURTESY OF SQUARE D COMPANY)

Fig. 3-31. An indoor service entrance panel. (COURTESY OF SQUARE D COMPANY)

4

Tools
for the Job

MOST HOMEOWNERS ALREADY HAVE A BASIC SET OF TOOLS; HOWEVER, if you are going to do much electrical work, consider investing in a few specialized tools. Remember, it is important to buy quality tools. A poorly made tool is never a bargain.

Some of the brands to look for include Channellock, Craftsman, Crescent, Klein, Lufkin, Snap-On and Stanley. One of the best lineman's pliers or side cutters available is made by Klein, while Crescent and Klein both make excellent diagonal cutters. You might need to go to an electrical wholesale supply to buy Klein lineman's pliers and diagonal cutters. Tongue-and-groove pliers are made by a number of manufacturers and tend to look very similar, but Channellock seems to be the preferred brand.

Lineman's pliers (Fig. 4-1) are probably the basic tool for electricians. Grooved jaws hold wires firmly and are useful for twisting bare wires together. Just behind the jaws are wire cutters. Lineman's pliers usually are equipped with insulated handles, which provide some measure of safety; however, it is still your responsibility to make certain the circuit you are working on is dead.

True insulated grips are sold separately and must be installed by the user. You can soften them by heating in boiling water, then while the grips are pliable, force them on the handles by tapping with a plastic hammer or a block of wood. Plastic grips installed by the manufacturers are put on by dipping the handles in the plastic solution. The plastic coating is thin and easily broken and its purpose is for comfort only and not for electrical insulation.

Another type of pliers used almost as much as lineman's pliers are diagonal cutting pliers (Fig. 4-1), more commonly known as *dikes*. They are used for cutting wires up to a #6 diameter.

Tongue-and-groove pliers (Fig. 4-1) are very useful when working with conduit. They come in a variety of sizes, from 6½-inch to the large 20-inch size that can grip pipes up to 5½ inches in diameter.

Needle-nose pliers (Fig. 4-1) come in a number of sizes, but the 6- or 8-inch size is about right for looping wire around screw terminals.

Crimping tools (Fig. 4-1) receive a lot of use, so a quality tool is important.

Wire strippers (Fig. 4-1) are available in a variety of forms. The simplest comes in a two-piece scissors-type arrangement. You might

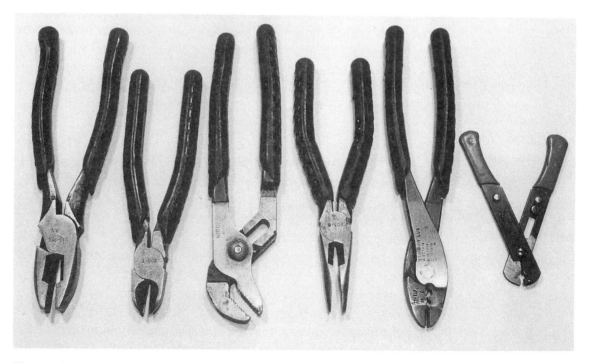

Fig. 4-1. Basic electrical tools. From left to right, lineman's pliers, diagonal cutting pliers, tongue-and-groove pliers, needle nose pliers, crimping tool, and wire strippers.

need a little practice before you can strip the insulation without nicking the wire, but this can be mastered after a few tries.

A cable ripper (Fig. 4-2) is an inexpensive tool that speeds up the job of stripping the outer insulation from the flat two-wire nonmetallic cable used in most homes. This simple tool has a small triangular blade positioned in such a way that when the tool is pressed around a cable and then pulled, it will make a slit in the insulation, allowing the outer cover to be easily peeled off. A combination wire stripper and crimper is also available (Fig. 4-2).

You will need some sort of testing device to check for voltage and continuity. A simple tool to check for voltage is the neon voltage tester (Fig. 4-2). If you touch one probe to a hot wire or terminal and the other to the neutral or ground, the lamp will light if the circuit is hot.

An inexpensive volt-ohmmeter (Fig. 4-3), available at any electronics supply store, will check for both voltage and continuity.

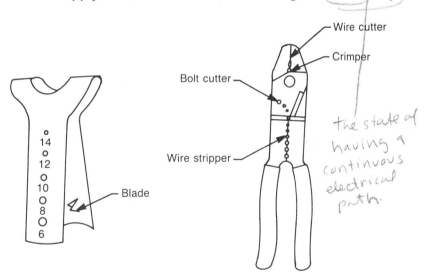

the state of having a continuous electrical path.

Fig. 4-2. Cable ripper, wire stripper/crimping tool, and a neon voltage tester.

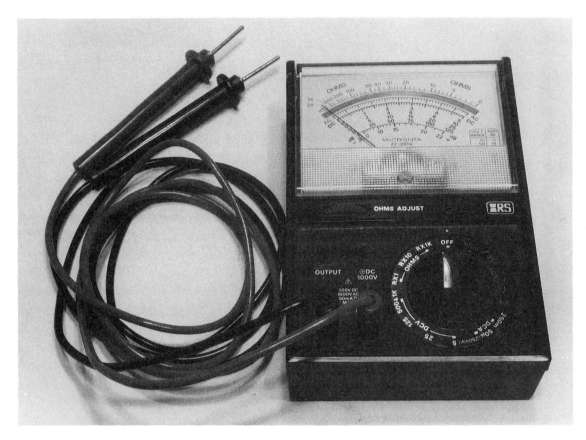

Fig. 4-3. Volt-ohmmeter.

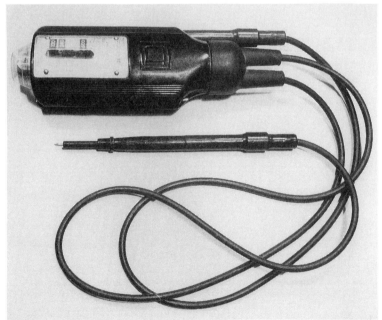

Fig. 4-4. Wigginton "Wiggy" voltage tester.

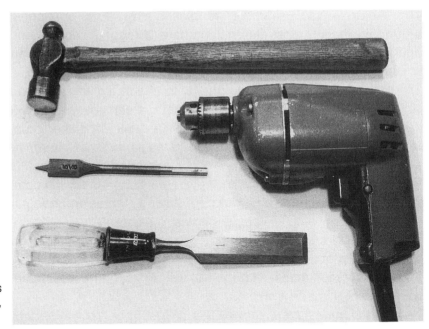

Fig. 4-5. Woodworking tools include a hammer, a chisel, a drill, and a wood bit.

Fig. 4-6. Tape measure and a screwdriver with a ⅜-inch blade.

Fig. 4-7. Screwdriver tips should be kept square.

This meter is also useful for troubleshooting faulty appliances around the home. Most electricians have a rugged, reliable voltage tester such as the one shown in Fig. 4-4. These testers are specifically designed to read 120-240-480 and 600 volts.

When using any test equipment to check for voltage, work slowly and carefully. Don't touch any metal. Keep your hands well back on the insulated area of the probes. If a carelessly placed probe is touching a hot wire or terminal and at the same time is touching a neutral or ground, a short will occur and sparks will surely fly.

Woodworking tools such as those shown in Fig. 4-5 will come in handy. You will need a hammer for driving staples and hanging boxes and a chisel for notching studs and plaster. An electric drill with about a ⅝-inch wood bit and a long heavy-duty extension cord will save a lot of time when you are drilling holes in joists or studs for wiring. A tape measure (Fig. 4-6) will be useful for positioning wall receptacles and switches.

Most homes have a screwdriver (Fig. 4-6) or two lying about, but one with a blade about ⅜ inch wide will handle most terminal screws. The important thing is to use a blade with the correct thickness for the size of the slot in the screw. Using the wrong size of screwdriver is the major cause of screws being damaged. Another point to remember about screwdrivers is that the tip should be squared (Fig. 4-7). Screwdrivers have a tendency to be abused

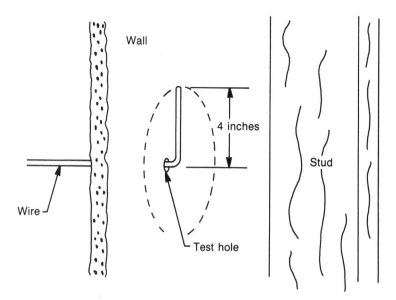

Fig. 4-8. Checking for obstructions.

and are often used as chisels and wedges. If the tip becomes rounded, it is no longer usable as a screwdriver. Use a file or grinder to keep the tip squared off.

Try to keep tools located in one place—a utility room or garage, perhaps. This way you'll be able to find them when they're needed. It can be very frustrating on a 20-minute job to spend an hour looking for the tool you need.

Before cutting a large hole in a wall, you should probe the area for obstructions. Drill a small test hole at the desired box location, then bend a 10-inch length of stiff wire into an L shape. Next push it through the hole and rotate the bend (Fig. 4-8). If the wire encounters an obstruction, try another test hole a few inches from the first position until you find a clear space.

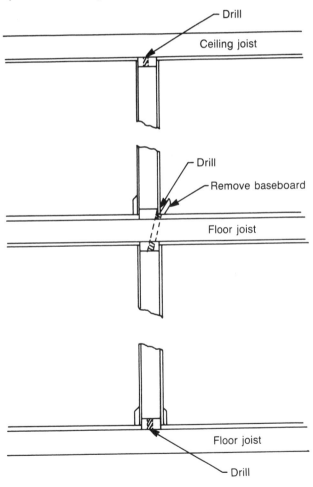

Fig. 4-9. Holes drilled from different elevations to route cable.

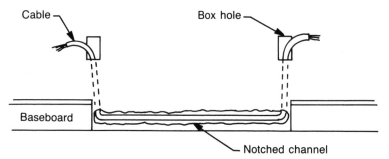

Fig. 4-10. Cable running behind baseboards.

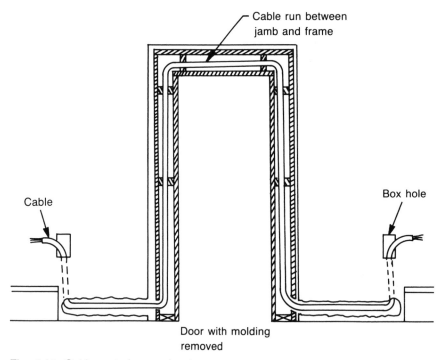

Fig. 4-11. Cable routed around a doorway.

You might have to use a lot of ingenuity for some wiring projects, but normally you can fish cable either down from the attic or up through a basement or crawl space. Figure 4-9 illustrates three different options for drilling holes to bring cable into wall cavities. One choice for running cable from one receptacle outlet to another is to remove a section of the baseboard and make a channel for the cable. The channel will be hidden when the baseboard is put back in place (Fig. 4-10).

You also can route cable around doorways (Fig. 4-11). First remove the molding from the door frame and whatever section of baseboard necessary to reach the box locations. Next run the cable between the jamb and the frame. Spacers are used inside the frame, so you will have to cut notches in them to allow room for the cable.

It is important to work slowly and carefully. Molding splits easily. Use a wide putty knife or chisel to pry the molding away from the wall. When the channel is being notched, keep in mind that the cable must be installed at least 1¼ inches from the finished surface; otherwise you must use conduit or a ¹⁄₁₆-inch metal plate to protect the cable.

5

Wiring Outlets
and
Overhead Fixtures

BEFORE BEGINNING ANY PROJECT, DO SOME PLANNING. DECIDE WHAT you want to add and where you want to put it. It's a good idea to put your plan on paper along with a list of materials needed. Determine your source of power. It may be the service entrance panel, or you might need to tap into an existing circuit at a light, switchbox, or receptacle. To do so, determine the capacity of that circuit. Will it handle the additional load? Then decide what size and type of wire to use.

MOUNTING BOXES

One of the first steps in installing an electrical circuit is to mount the boxes for the switches, receptacles, and lights (Fig. 5-1). These boxes may be metal, or nonmetallic, such as plastic (see Fig. 5-2). The nonmetallic boxes are popular because the single-gang switchboxes don't require cable clamps, and they do not need to be grounded. This saves time on installation, but both types have their advantages.

MOUNTING SWITCHES AND RECEPTACLES

You can usually mount switches and receptacles on studs, but often you must locate ceiling fixtures between joists. Adjustable

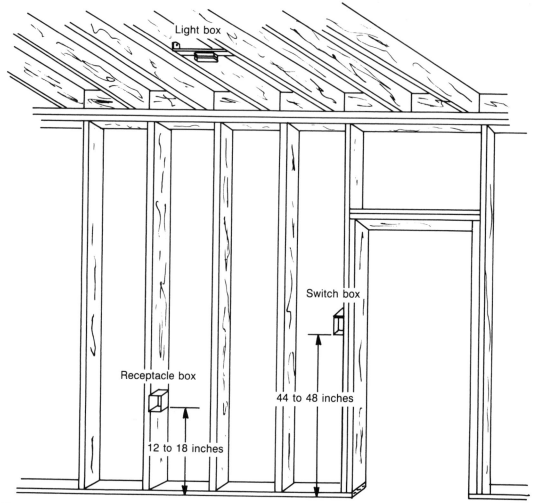

Fig. 5-1. Boxes mounted prior to running wire.

hangers are available (Figs. 5-3 and 5-4). It is important to know that there is a limit on the number of wires a box will legally hold (Table 5-1). This limit is based on the volume in cubic inches of the box. There must be enough free space for all the conductors that will be enclosed. The same boxes are available in different depths. It only costs a little more to use the deeper boxes, but the extra room makes it easier to fold the wires in place when installing the switches, light fixtures, and receptacles.

Mount receptacles 12 to 18 inches above the floor, and the code requires them to be close enough together so that no point along the wall will be farther than 6 feet from an outlet.

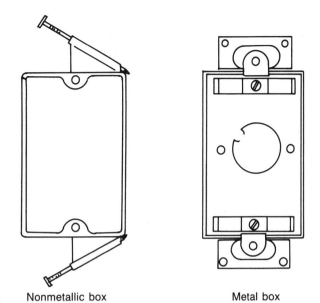

Fig. 5-2. Two types of electrical boxes.   Nonmetallic box   Metal box

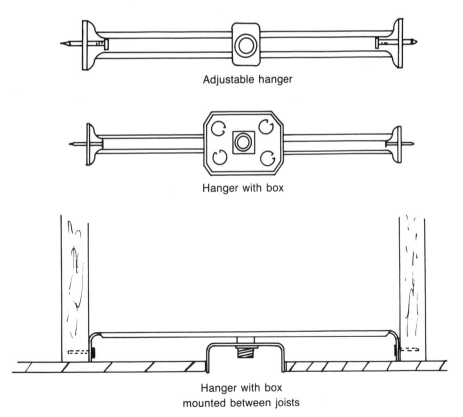

Adjustable hanger

Hanger with box

Hanger with box
mounted between joists

Fig. 5-3. Adjustable hangers are used to mount boxes between joists.

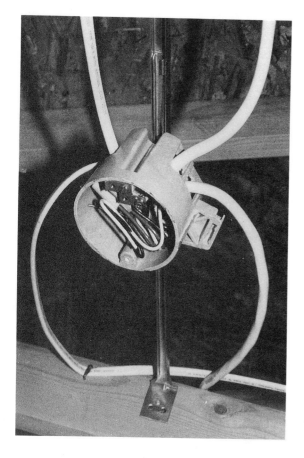

Fig. 5-4. Ceiling fixture supported by adjustable hanger.

Position switch boxes 44 to 48 inches from the floor and locate them for easy access when the door is opened. They should always be on the opposite side from the hinges, and you should install them so that the front edges of the box will be flush with the finished wall or ceiling. After the boxes are mounted, drill holes or cut notches, and run the cable from the power source to the boxes, but don't connect to the power until you have wired in the other devices. Avoid twists or kinks in the cable, and when possible, route cable along the sides of structural members such as studs, joists, and rafters (Fig. 5-5).

When estimating cable length, draw a rough sketch of the cable route. Add the distance between boxes and then the distance from the ceiling or floor to each box. Then add about 2 feet for each box to cover unforeseen obstacles and to make connections. When drilling holes, use the smallest bit possible (About ⅝ or ¾ inch) to avoid weakening the wood structure. Drill in the center of the studs. If

Table 5-1. Various Box Sizes and the Number of Wires They Will Hold.

Type of Box	Box Size in Inches		Maximum Number of Wires			
			No. 14	No. 12	No. 10	No. 8
Outlet Box:	4 × 1-¼	Round	6	5	5	4
	4 × 1-½	or	7	6	6	5
	4 × 2-⅛	Octagonal	10	9	8	7
	4 × 1-¼	Square	9	8	7	6
	4 × 1-½	Square	10	9	8	7
	4 × 2-⅛	Square	15	13	12	10
	4-11⁄16 × 1-¼	Square	12	11	10	8
	4-11⁄16 × 1-½	Square	14	13	11	9
	4-11⁄16 × 2-⅛	Square	21	18	16	14
Switch Box:	3 × 2 × 1-½		3	3	3	2
	3 × 2 × 2		5	4	4	3
	3 × 2 × 2-¼		5	4	4	3
	3 × 2 × 2-½		6	5	5	4
	3 × 2 × 2-¾		7	6	5	4
	3 × 2 × 3-½		9	8	7	6
Handy Box:	4 × 2-⅛ × 1-½		5	4	4	3
	4 × 2-⅛ × 1-⅞		6	5	5	4
	4 × 2-⅛ × 2-⅛		7	6	5	4

the hole is less than 1¼ inches from the edge, you must install a metal plate to protect the cable from nails. Secure cables at least once every 4 ½ feet and within 12 inches of each metal box and 8 inches of each plastic box. You may use staples (Fig. 5-6), but take care not to damage the wires.

If you use a metal box, you must secure the cable to the box by a cable clamp or cable connector (Fig. 5-7). Metal boxes must also be grounded (Figs. 5-8 and 5-9). The cable need not be secured to single-gang plastic switchboxes, and because they do not conduct electricity, they are not grounded.

Cables are made up of a bundle of insulated and bare wires inside a cover of insulation. Before you can make the connections to a switch or other device, you must remove the outer insulation and cut away any separation material. Then cut the wires to a manageable length and strip the insulation from the ends to make the connections.

If you are installing a flat two-wire (with or without ground), nonmetallic sheathed cable, use a cable ripper. Just slide the cable ripper over the cable and up inside the box, then squeeze the handles together and slowly pull down off the end of the cable to

Fig. 5-5. Cables routed along joists and through attic.

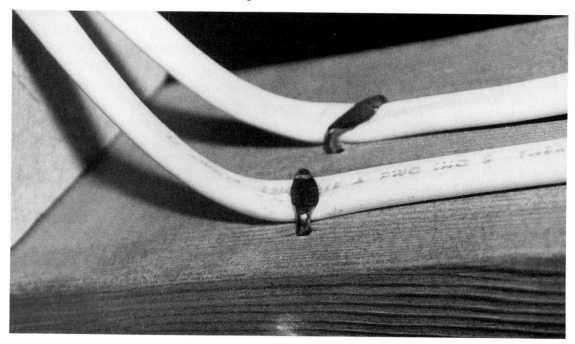

Fig. 5-6. Cables held in place with staples driven into wood members.

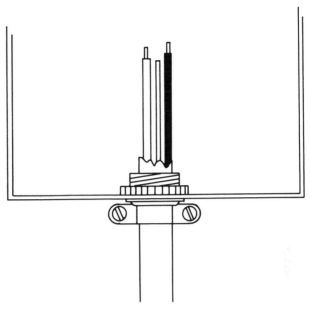

Fig. 5-7. A cable clamp holds the cable securely as well as providing protection against the rough edges of the hole.

cut or score the outer insulation. Next peel back this cover, and cut off the insulation and any separation material with a pair of dikes.

If you are installing the round, three-wire cable, such as the kind used when wiring three-way switches, use a pocket knife or utility knife, and cut it following the twisted rotation of the wires. Try to cut between the wires without cutting their insulation. Never cut any cable using your hand or your knee for a workbench. Use a flat board or a nearby wall surface. Leave about 6 or 8 inches of wire extending from the box to make up the connection, and fold the wire back in the box (Figs. 5-10 and 5-11). When stripping the insulation from the ends, be careful not to nick the wire. Wire strippers work well for sizes up to #10. From #8 wire on up, use a pocket knife in the same way you would to sharpen a pencil (Fig. 5-12).

When connecting a wire to a screw terminal, strip away about ½ to ¾ inch of the insulation, and form a half loop in the uninsulated end. Next hook the bare wire clockwise around the screw. This way, when you tighten the screw, the loop will close around it. If the wire is hooked counterclockwise, the loop will open and be forced from underneath the screw head. Don't allow any of the insulated wire

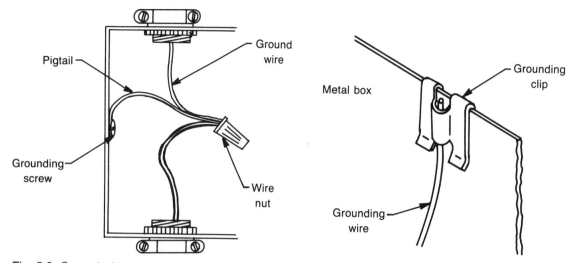

Fig. 5-8. Ground wire attached to a metal box by a grounding screw.

Fig. 5-9. A grounding clip is sometimes used to connect the ground wire to a metal box.

to be clamped by the screw head, and try not to let more than about ¹⁄₁₆ inch of bare wire extend past the connection. Limit screw terminals to one wire. If you need to connect more than one wire to one terminal, use a *pigtail splice*, (Fig. 5-13), which is simply three or more wires connected together and one of the wires (the pigtail) is connected to the terminal.

When replacing or installing switches, read the information on the switch to be sure it has the same or higher current and voltage capacity for that circuit. Today most switches and receptacles have both screw terminals and push-in connectors. Switches found in the home tend to be single-pole, three-way and four-way types. Single-pole switches have a top and bottom and are mounted so the toggle is up when the switch is on and down when the switch is off. They also have two terminals with the same color. Only hot wires are connected to switches, so with a single-pole switch it doesn't make any difference which wire is connected to which terminal. However, if the wires run to the light first, thus placing the switch at the end of the run (Fig. 5-14), be sure to mark the white wire with black tape to show it is a hot wire. To wire the switch when it is in the middle of the run (Fig. 5-15), connect the two neutral (white) wires together. Make the ground (bare) wire connections, and connect the two remaining black wires to the switch.

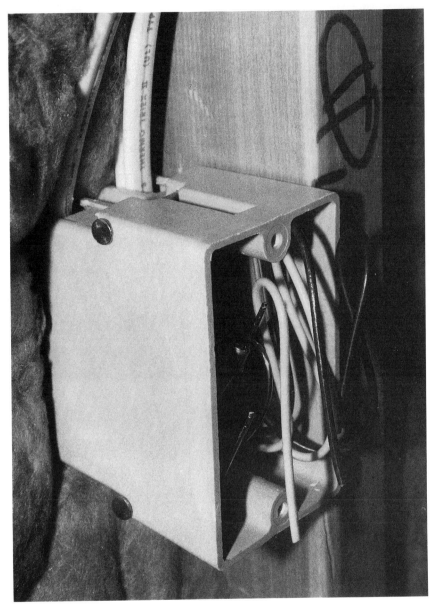

Fig. 5-10. Nonmetallic box mounted and wired for receptacle. Notice marking on stud indicating receptacle.

Receptacles are also easy to install (Fig. 5-16). Strip the ends of the wires using the molded gauge on the back of the receptacle, then make ground connections. You might have to make up a pigtail if the receptacle is in the middle of the run (Figs. 5-17 and 5-18). Insert the white wires in the holes in the back, on the side with the

Fig. 5-11. Nonmetallic box mounted and wired for three switches.

Fig. 5-12. Number 8 wire and larger is normally stripped as you would when sharpening a pencil.

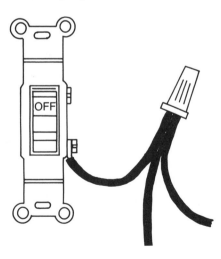

Fig. 5-13. A pigtail splice.

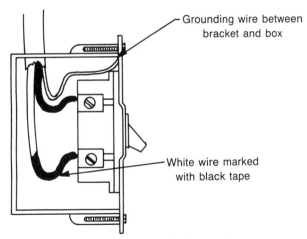

Grounding wire between bracket and box

White wire marked with black tape

Fig. 5-14. A switch installed at the end of a circuit.

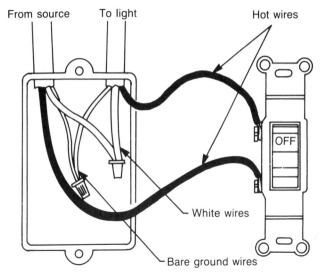

From source   To light   Hot wires

OFF

White wires

Bare ground wires

Fig. 5-15. A switch installed in the middle of a circuit.

silver-colored screws. Insert the black (hot) wires in the opposite holes in the back. They should be on the same side as the brass screws.

Most of the circuits in a home will be 15- or 20-amp general-purpose circuits. Major appliances, however, require their own individual circuit. Which appliance is considered major might be confusing, so check with your local building inspector first. Generally, your installations will meet the code if you consider the following as major appliances: water heater, kitchen range, clothes dryer, washer, dishwasher, garbage disposal, and any device that

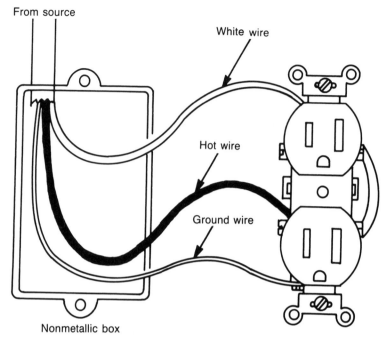

Fig. 5-16. A receptacle installed at the end of a circuit.

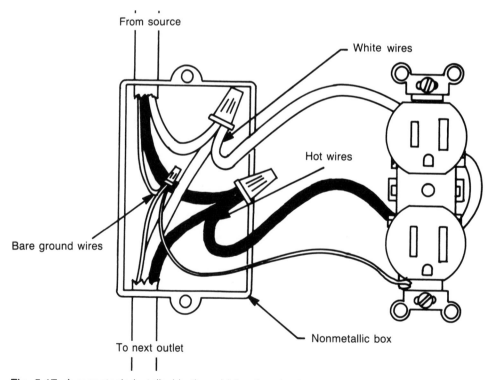

Fig. 5-17. A receptacle installed in the middle of a circuit.

Fig. 5-18. Receptacle installed in box.

Service switch?

employs an automatically started motor, such as a pump, oil burner, or air conditioner. The code further requires that each appliance be provided with proper overcurrent protection, as well as some means for completely disconnecting the appliance from the circuit.

An electric range uses both 120 and 240 volts. When the burner is set on low heat, it uses 120 volts, but when the burner is turned to the high setting, it uses 240 volts. The code states that the receptacle should be located within 6 feet of the location of the

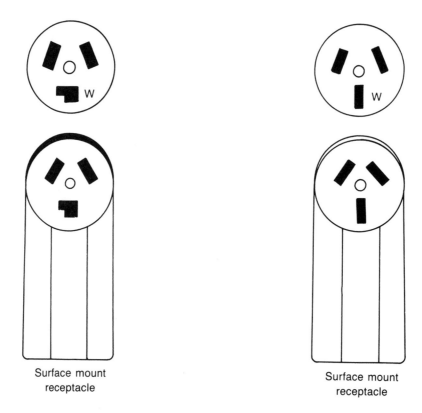

Fig. 5-19. Dryer receptacle: 30 amps, 125/250 volts.   **Fig. 5-20.** Range receptacle: 50 amps,125/250 volts.

appliance. In normal practice, however, locations are much closer. Some areas require range receptacles to be installed in metal boxes. Always consult the local building inspector.

A typical circuit for some appliances would be: dishwasher and garbage disposal, 120-volts and 20 amps wired with #12 copper wire; an electric dryer, 120/240-volts and 30 amps wired with #10 copper wire; an electric range, 120/240-volts and 50 amps wired with #6 copper wire. When the wires feeding the appliance come directly from the service-entrance panel, three conductors (usually black, red, and white) are necessary, but if they are fed from a subpanel, a fourth (ground wire) will be required.

A dryer will use a 30-amp 120/240-volt receptacle such as shown in Fig. 5-19. Note the opening marked *W*. It is for the white wire, while the other two are for the hot wires.

An electric range will use a 50-amp 120/240-volt receptacle like the one illustrated in Fig. 5-20. In this case, two openings are angled, while one is straight. The two angled openings are for the hot wires,

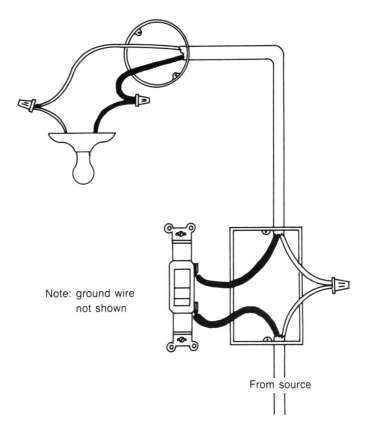

Fig. 5-21. Air-conditioner receptacle: 30 amps, 250 volts.

G

and the straight one is for the white wire. These receptacles are for 120/240-volt circuits.

A straight 240-volt circuit, such as one used for an air conditioner, will use a receptacle like the one shown in Fig. 5-21. Central air conditioners are usually fed from a subpanel located near the unit.

LIGHTS

Light fixtures come already wired, usually one black wire and one white wire, so it becomes a simple matter to connect the black

Note: ground wire not shown

From source

Fig. 5-22. A light at the end of a circuit where the power comes to the switch first.

fixture wire to the hot wire and the white fixture wire to the white neutral wire (Figs. 5-22 and 5-23). You can usually attach ceiling fixtures directly to an outlet box (Fig. 5-24); however, some are supported by a threaded stud, which is supplied with the fixture. When the fixture is to be flush with the surface make sure you allow for the ceiling thickness (Fig. 5-25).

CEILING FANS

Wire in ceiling fans just like you would a light fixture. Because of the added weight, however, you will need to mount the box securely. This usually means attaching the box directly to a wooden joist or brace between the joists (Fig. 5-26). You can use a threaded eye bolt to attach the fan to the wood member. Screw the bolt through an opening in the back of the box and into the wooden support. As with any fixture or device, do not place any strain on any of the wires. Their only purpose is to conduct electricity.

Safeguards

Manufacturers normally include detailed instructions with their fans. Thomas Industries, Inc., manufacturer of Lafayette fans, includes the following information in its installation guide.*

Important Safeguards

1. To ensure a proper installation, read the instructions carefully and review the diagrams thoroughly before handling your fan.

2. Proper care must be taken when working with electrical wiring. All electricity must be turned off at the main power box before installing. Failure to observe this procedure may result in possible electrical shock. All electrical wiring connections must be made in accordance with all local codes, ordinances and the National Electrical Code. If you are not familiar with the methods used to install and hook up electrical wiring, you should secure the services of a qualified electrician.

3. Measure your intended hanging site very carefully. Allow enough space so that the rotating fan blades will not come in contact with any obstacle.

---

*Reprinted by permission of Thomas Industries, Inc.

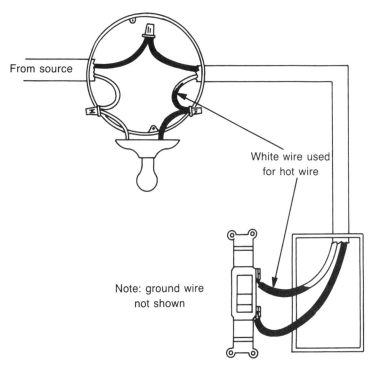

From source

White wire used
for hot wire

Note: ground wire
not shown

Fig. 5-23. A light installed where the power comes to the light first.

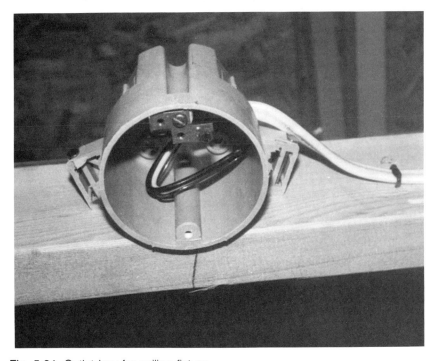

Fig. 5-24. Outlet box for ceiling fixture.

Fig. 5-25. Ceiling fixture mounted flush with surface.

Fig. 5-26. Wooden support is used to support heavy fixtures and ceiling fans.

4. When mounting the fan to an outlet box, a standard 4″ × 2⅛″ metal octagon electrical box must be used. The electrical box must be firmly secured to a joist or reinforced to be able to support the hanging and moving weight of the fan (at least 40 lbs.). Failure to properly secure the box will result in an improper operation or possible damage. Caution: Do not use a plastic electrical box. It will not adequately support the fan.

5. Do not attach blades until the motor housing has been hung and is properly secure. (To prevent marring or scratching of finish, the motor housing should be kept in its carton until installation.) Make sure that all connections and screws are tightened securely to prevent the fan from falling, which could produce serious bodily harm and damage.

6. The fan can be operated immediately after installation. The fan ball bearings are adequately charged with proper type oil. Under normal operating conditions additional lubrication will not be necessary.

IMPORTANT NOTICE: The safeguards and operating instructions in this manual cannot fully cover all possible conditions and situations. To avoid any risk of injury, common sense and electrical safety are a definite must.

Assembly of Switch Housing With Motor

See Fig. 5-27. Before removing the motor from its lower Styrofoam base, place the empty Styrofoam top cover upside down. Place the motor on this cover to prevent inadvertent scratching or marring of the motor finish.

1. Join the terminal plug from the motor to the terminal plug from the switch housing. The two plugs must be joined in the correct direction so that the keys on both sides of one plug will fit completely into the two side slots of the other plug. (Each plug has color coded strips on its housing. When properly matched, the result is a correct alignment.)

2. Attach the switch housing cover to the base of the canopy and secure the three screws.

Assembly of Downrod with Motor

See Fig. 5-28. If you receive your ceiling fan with downrod and

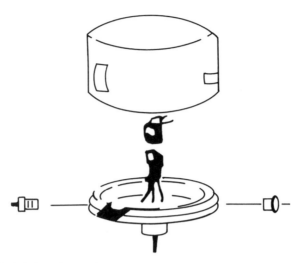

Fig. 5-27. Switch housing and motor. (COURTESY OF THOMAS INDUSTRIES, INC.)

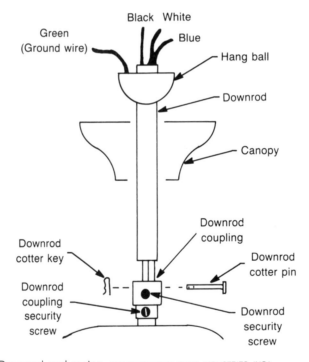

Fig. 5-28. Downrod and motor. (COURTESY OF THOMAS INDUSTRIES, INC.)

motor assembly already together, you can omit this assembly procedure.

1. Slip the canopy along the downrod so that it covers the hang ball. Feed the black and white power wires and blue light wire

through the downrod (start at lower end) until they emerge from the top.

2. Secure the downrod to the coupling properly to prevent possible separation when the fan is operated in reverse mode.

3. Tighten the downrod coupling security screw. Loosen the downrod security screw to allow for complete entry of downrod. Place the downrod into the downrod coupling and visually align the holes in both the downrod and downrod coupling. Then install the downrod cotter pin through the holes and insert the downrod cotter key. Secure by twisting the ends of the cotter key with a pair of pliers so removal is not possible.

Test the assembly by attempting to twist the downrod. It will hold when properly secure.

Assembling the Blades and Blade Holder

See Fig. 5-29.

1. Attach the blades to the blade holder by using screws and the metal washers as provided.

2. There are three screws per blade that are threaded through the washers, through the wood blade, and through the blade holder arm.

Note: You may be required to follow a different method of blade assembly if nuts are provided instead of flat washers. (See Fig. 5-29B.)

1. The method to assemble the blades to the blade holder is as above except for the use of lock spring washers and nuts.

2. The three screws per blade are threaded through the blade holder, through the wood blade, and tightened through the regular washer, spring washer, and finally tightened securely with the nut.

Fan Installation

See Fig. 5-30. Note: Before final installation, turn off electrical power at main power box.

1. Securely position the mounting bracket to the ceiling junction box by using the provided junction box screws and washers.

2. To avoid wobble or poor fan performance, be sure that the mounting bracket is securely tightened against the ceiling.

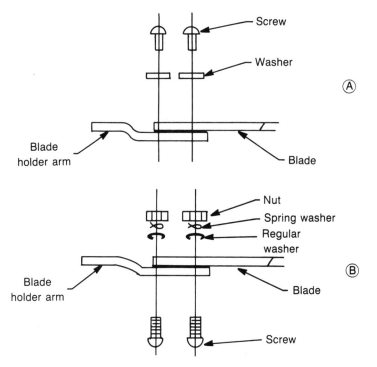

Fig. 5-29. Assembling the blades and blade holder. (A) Assembly using screws and washers. (B) Assembly using screws and nuts. (COURTESY OF THOMAS INDUSTRIES, INC.)

3. Lift the downrod carefully into the mounting bracket, then turn the convex of the hang ball cushion into the slot position. Be sure that the motor is firmly seated on the mounting bracket.

4. Tighten the downrod security screw. (Refer to Fig. 5-28.)

5. Use the provided wire nuts to wire connections together as indicated. (Refer to schematic section Fig. 5-31).

    a. Separate wires. Position green and white wires at one side and black and blue wires at the other side of junction box.

    b. Connect the black wire from fan assembly to (positive) black wire from the electrical power.

    c. Connect the white wire from the fan assembly to the (neutral) white wire from the electrical power.

    d. Should you desire the addition of a light kit, join the black and blue wires together and connect to the black (positive) wire from the electrical source.

    e. Connect the green wire (ground) from fan to house ground wire.

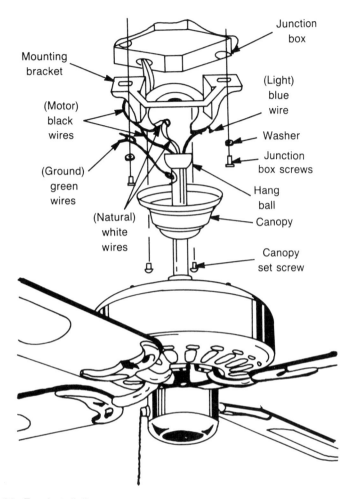

Mounting bracket

Junction box

(Light) blue wire

(Motor) black wires

Washer

Junction box screws

(Ground) green wires

Hang ball

Canopy

(Natural) white wires

Canopy set screw

Fig. 5-30. Fan installation. (COURTESY OF THOMAS INDUSTRIES, INC.)

    f. The joined connections (splices) should be turned upward and carefully pushed into junction box.

    6. Slide the canopy up the downrod until it is flush with the ceiling, and securely tighten the set screws in the canopy.

    7. Attach one blade holder arm to the fan housing using two blade-mounting screws. Rotate the motor by hand until the blade is opposite you, then attach the second blade and additional blades. (See Fig. 5-32A.)

    Under normal operating conditions the directional switch should be in the down position, which is best for summer operation. For the best results in the winter, the directional switch should be in the up position, which reverses the fan direction.

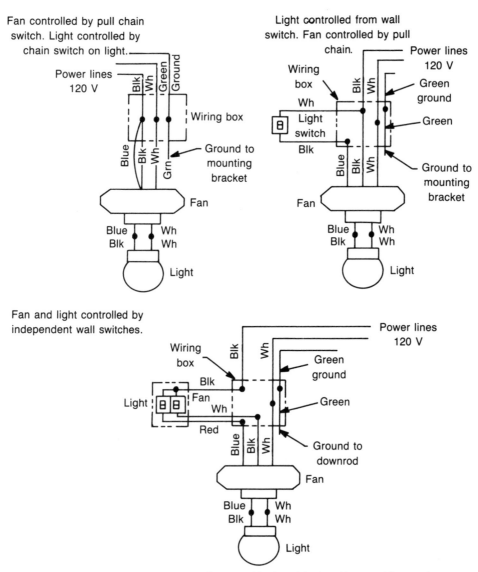

Fig. 5-31. Make electrical connections. Connect black to black, white to white, and green to green or bare wire (ground). Make sure that fan is properly grounded. Note: Light kit leads shall be suitably capped with a connector, secured with tape, and turned upward into the switch compartment. (COURTESY OF THOMAS INDUSTRIES, INC.)

Proper Balance for Your Fan

If your Lafayette ceiling fan exhibits a wobble during operation, you must check the mounting system to determine if it was improperly installed. The following procedures should alleviate the problem.

1. Be certain that the mounting canopy has not become warped by over-tightening the mounting screws. If this has happened, the mounting ball will rock in the canopy instead of being seated tightly. This can be corrected by loosening the screws, reseating the mounting ball, and then carefully retightening the mounting screws until just firm. In some cases, a gentle rocking of the ball in a circular motion will seat it tightly against an already distorted mounting canopy.

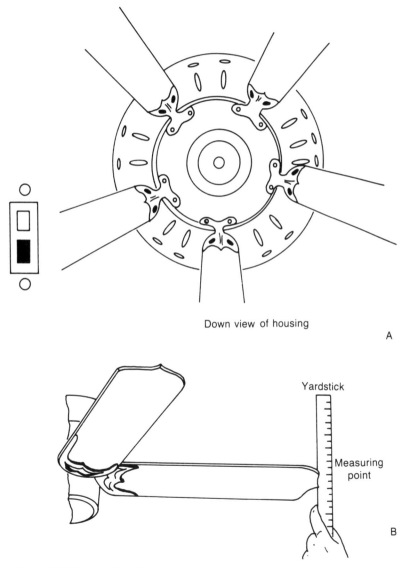

Down view of housing

A

Yardstick

Measuring point

B

Fig. 5-32. Reverse fan direction. (COURTESY OF THOMAS INDUSTRIES, INC.)

2. The Lafayette ceiling fan canopy must always be securely tightened against the ceiling to prevent a wobbly movement. This is accomplished during the installation by the uniform tightening of the mounting screws. It is important to make sure the ceiling junction box is sufficiently supported. If the box is not providing adequate support, then it must be additionally braced to avoid irregular motion and possible damage.

3. Be sure that the downrod assembly is securely tightened to prevent wobble. After any adjustment to the downrod, always check to make sure that the lockscrew is firmly tightened.

4. Check to be sure that all the fan blades are firmly tightened into the blade holders and that blade holders are also properly attached to the flywheel.

5. Check to be sure that none of the blade holders is bent. This will result in a blade being out of position. This can be corrected by gently bending the blade holder back into the correct position.

6. A simple yardstick or ruler can be used to check for proper tracking (Fig. 5-32B). Position the yardstick vertically against the ceiling and even with the outer edge of the blade. Note the distance from the edge of the blade to the ceiling. Turn the blades slowly by hand to check the remaining blades. If properly aligned, the measurement from the ceiling should be the same for all of the blades. If a blade is slightly out of alignment, the blade holder can be gently bent either up or down to align with the other blades.

After the devices have been installed, make the connection to the power source and turn the power back on.

Electrical wiring should be kept simple and straightforward. Be neat; sloppy wiring can be dangerous. It is the first sign of an amateur at work and brings the effort under suspicion.

6

Thermostats, Door Chimes, and Outdoor Lighting

Today most homes are equipped with central heating and cooling systems. These systems are designed to efficiently maintain a comfortable temperature inside the residence. The heart of the comfort-control center is a sensing device we know as a *thermostat*. The thermostat is simply a switch that responds directly to a change in the temperature and automatically turns heating and air-conditioning units on or off. They are usually operated jointly with a thermometer set to the desired temperature.

Typical home thermostats operate on the principle that two different metals expand in different amounts under the same temperature change. These thermostats are made up of two strips of different metals bonded together, often in the shape of a coiled spring. If the surrounding temperature changes, the different metals expand or contract unequally. As a consequence, the bonded strip slowly bends in the form of an arc. This bending is the switching action that makes or breaks the electrical connection, turning the desired unit on or off (Fig. 6-1).

THERMOSTAT CIRCUIT HOOKUP

Some thermostats operate on line voltage (120 volts ac), but more often, the heating or air-conditioning unit is equipped with a

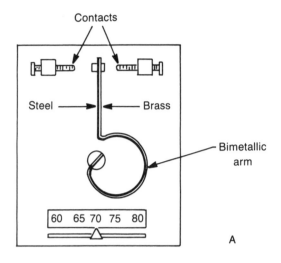

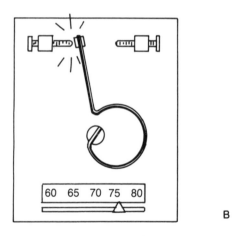

Fig. 6-1. Bimetallic thermostat, (A) Contacts open. (B) Heat causes brass to expand faster than steel, which moves arm and closes contacts.

low-voltage transformer (Fig. 6-2). This transformer lowers the control voltage to about 24 volts ac.

Low-voltage thermostat cable may have five or six color-coded wires, depending on the units and the thermostats. Connections are clearly marked on both the unit and the back of the thermostat (Fig. 6-3) with letters such as *W*, *Y*, *R*, and *G*. Typically the *W* connection will affect the heating relay. The *Y* controls the cooling contactor. The *R* will connect to the manual fan ON, and the *G* will operate the fan automatically in either heat or cool.

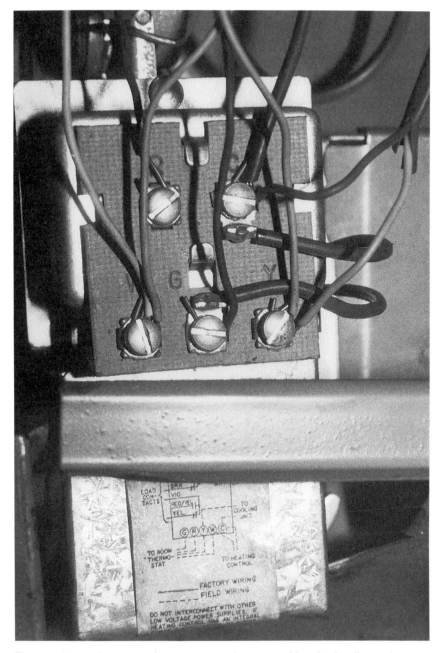

Fig. 6-2. Low-voltage transformer mounted on central heating/cooling unit.

Most thermostats have a small mercury-filled glass bulb attached to the bonded strip (Fig. 6-4). This mercury makes the electrical connections, so the base of the thermostat must be mounted level.

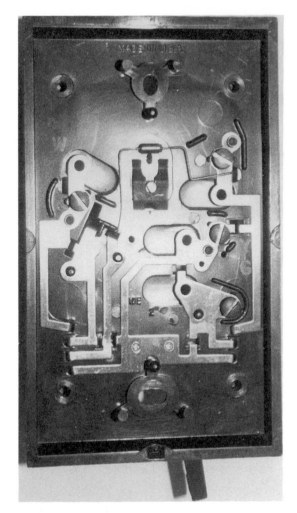

Fig. 6-3. Back view of thermostat.

Fig. 6-4. Mercury in glass bulb serves as switch.

Use a small level, and mark the spots for the mounting holes with a pencil. Use a small drill bit to make the holes. (In drywall, screw anchors may be necessary). Next mount the base on the wall, and attach the cover (Fig. 6-5). Move the control to the OFF position, then make the connections at the unit, turn the power back on, and set the thermostat.

DOOR CHIME CIRCUIT AND TRANSFORMER

Doorbells may ring or chime, but they all operate from a low-voltage circuit. The low voltage is provided by a transformer. The transformer steps the 120-volt line voltage down to the voltage

Fig. 6-5. Thermostat
mounted in place.

Fig. 6-6. Residential door chime.

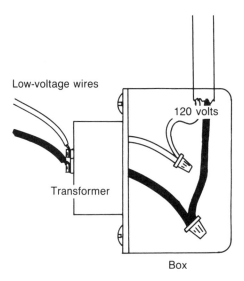

Low-voltage wires

120 volts

Transformer

Box

Fig. 6-7. Wire connections to a transformer.

required by the doorbell or chime. For doorbells and buzzers, this may be about 6 to 10 volts. Door chimes (Fig.6-6), however, might require a transformer that will deliver about 20 volts (Figs. 6-7 and 6-8). If you replace a bell with a chime, the sound might not be loud enough; you probably will have to replace the transformer as well. In low-voltage circuits, there normally is little danger of shock and the installation is simple. Think of the low-voltage, or output, terminals on the transformer as the power source for the circuit, the bell or chimes as the load (Fig. 6-9), and the button as the switch (Fig. 6-10). Figure 6-11 is a diagram showing a single-button arrangement.

In some installations, one doorbell is not enough and a second signal is needed for a back or side door. This is easily arranged by adding another button (Fig. 6-12). Most door chimes are already equipped so that two chimes will sound if the front button is pressed and one chime will sound if the rear button is used (Fig. 6-13). The terminals on the chimes are labeled front and rear for convenience of installation.

Some local codes state that transformers must not be located in attic spaces. Check with the building inspector before making any installations. Transformers can be found in closets, utility rooms, and garages; however, the local code may state that the transformers must be readily accessible and must not be positioned near any combustible material that might create a hidden fire dan-

Fig. 6-8. Transformer mounted
on a metal box.

ger. The 120-volt supply could come from almost any 120-volt circuit
not controlled by a switch. Before replacing a faulty chime, check
it out. Often a chime will not sound because it is gummed up with
dirt and lint. In troubleshooting doorbell circuits, you normally need
to have the power source connected, but kill the power if any work
on the input (120-volt) side of the transformer is to be done.

   If there are 120 volts going into the transformer, check the low-
voltage side. You should find 10 to 20 volts, depending on the

Fig. 6-9. Chime striker mechanism.

Fig. 6-10. Front door button.

transformer. Have someone press the button by the door while you listen to the chimes. If you hear a humming sound, chances are it's okay. The striker shaft may be binding and need cleaning. If not, there is probably a break in the circuit or the button could be bad.

To check out the button, carefully remove the cover (Fig. 6-14), and remove the mounting screws (Fig. 6-15). Next simply jumper across or disconnect the two wires and touch them together. If this sounds the chime, replace or repair the button (Fig. 6-16). Often just cleaning the striker mechanism or tightening a few electrical connections will bring back the musical voice to a silent door chime.

YARD LIGHTS

Yard lights will enhance the beauty of any landscape, as well as provide safe travel down steps and walkways. It further offers additional security against intruders because few burglars will en-

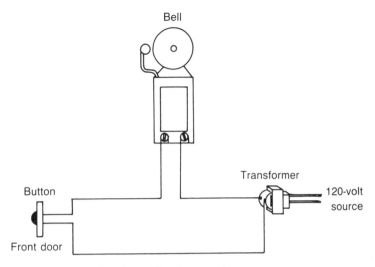

Fig. 6-11. Illustration showing wiring for one door.

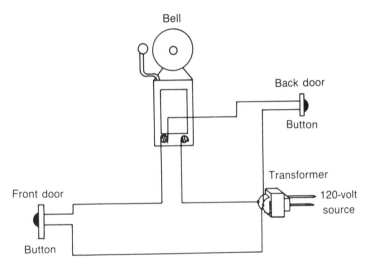

Fig. 6-12. Illustration showing wiring for a front and back door.

ter a well-lighted area. If you want to install a number of lights, a 120-volt system could become expensive, but by using low voltage, you will make the installation very simple and affordable. By using 12 volts, you can install the wiring more easily, and you won't have the shock hazards of a 120-volt system.

A transformer (Fig. 6-17), usually weatherproof and fitted with a timer, is used to step down the residential 120 volts to a safe 12 volts. This also eliminates the need for the conduit and weatherproof boxes normally associated with outdoor wiring.

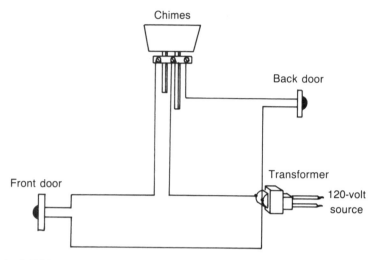

Fig. 6-13. Wiring arrangement where the front door sounds two chimes and the back door sounds one.

A weatherproof receptacle will be needed for a power source. Often you will need to install the receptacle to satisfy the needs of a particular location. They usually can be installed behind an existing receptacle inside the house. If the circuit is not already protected, use a ground-fault receptacle with a weatherproof box and cover.

Mount the transformer near the receptacle. Connect the wires supplying power to the lights to the low-voltage side of the transformer, and run the wire to the desired location. It's a good idea to bury the wires a few inches. Try to locate them where they are not likely to be damaged accidentally from digging later on. Each set of lights will have about 50 feet of wire.

Installing the lights is usually just a matter of pushing the pointed stands into the ground. Try to imagine the effects as you position each light. You probably will need to relocate some after you have tried them at night in order to achieve the desired effects.

A variety of light styles is available to satisfy different situations. Tier lights (Fig. 6-18) can provide illumination along flower borders such as sidewalks and driveways, while floodlights (Fig. 6-19) hidden by shrubbery close to walls can bring out architectural features. Colored filters are usually available seasonal or festive moods.

Connecting the wires to the lights is usually accomplished by simply pressing a clip from the light down over the wire (Fig. 6-20).

Fig. 6-14. Front door button with cover removed.

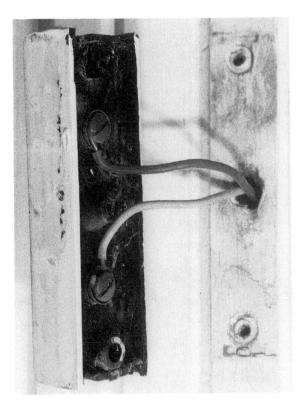

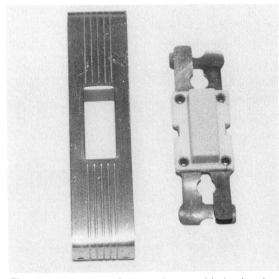

Fig. 6-15. (left) Mounting screws removed, exposing wire connections.

Fig. 6-16. (above) Button disassembled, showing electrical contacts.

Fig. 6-17. A weatherproof timer.

Fig. 6-18. Tier lamp.

Fig. 6-19. Floodlights softly illuminate wide areas.

Fig. 6-20. Tier lamp showing wire connection.

Two metal prongs inside the clip pierce the insulation and make the connection to the wires inside. Some assembly of the lights will probably be necessary, but this is easily accomplished with a screwdriver. Complete instructions are provided with each set of lights.

7

Upgrading an Existing Service

IF YOUR HOME'S ELECTRICAL WIRING IS ADEQUATE FOR YOU NEEDS, you might not want to run more wiring. However, you can upgrade the service and safety of your home without tearing out walls, by simply installing smoke detectors, dimmer or three-way switches, timers, and other devices.

SMOKE DETECTORS

More and more city ordinances are requiring smoke detectors to be installed in new construction. New homes in some areas must have a smoke detector installed in each bedroom, as well as one in the kitchen and one at the stairway. These home fire-alarm devices are about the size of a small light fixture (Fig. 7-1). They have been credited with saving many lives and sometimes property by sounding an alarm early enough to allow the occupants to escape and alert the fire department.

There are three different types of models on the market that are designed for homes: ionization units, photoelectric units and more expensive units combining both principles.

Ionization models respond 20 to 30 seconds faster than photoelectric ones to fast-burning fires, such as those fed by burning paper or flammable liquids. These few seconds might be critical in this type of fire.

The photoelectric detectors, however, respond sometimes 20 minutes earlier to smoldering fires such as those caused by a cigarette dropped on a mattress or on upholstered furniture. This type of fire tends to happen more often and probably should be considered a greater danger.

Logic would dictate that a unit that combined both principles would be the best choice. However, such logic might be subject to question for they are usually much more expensive and sometimes have low levels of performance in comparison.

It is highly unlikely anyone will know in advance, what type of fire will occur. It might be better to consider installing both an ionization and photoelectric detector.

You could locate the ionization model in a hallway near the bedrooms. Studies seem to indicate that most fires occur somewhere between 10:00 P.M. and 6:00 A.M.

You should place a photoelectric detector near the living area where it would detect smoke from a smoldering couch. Use one in the bedroom if anyone still smokes in bed.

Most smoke detectors can be mounted on the ceiling or the wall. The ideal location would be in the middle of the ceiling. Wall mounted detectors should be between 6 to 12 inches from the ceiling, never more than 12 inches from the ceiling and never in any corners. Try to mount the detector out of the path of ventilation where the air flows past the device faster than in other areas of the room.

Smoke detectors are available in battery operated or direct-wired models. The direct-wired detectors are permanently wired into one of the house circuits (Fig. 7-2), such as a bathroom circuit, where it will readily be noticed if the breaker is tripped. The battery-operated models operate independently of the home electrical system, which might be an advantage if the home's supply is interrupted for some reason. These units normally are powered by a single, nine-volt alkaline battery, which should be replaced every year (Fig. 7-3). Most models will beep periodically to signal that the battery has become weak. These inexpensive life saving devices are rapidly becoming necessities in today's homes.

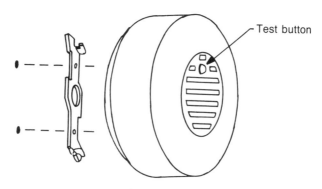

Fig. 7-1. An inexpensive smoke detector.

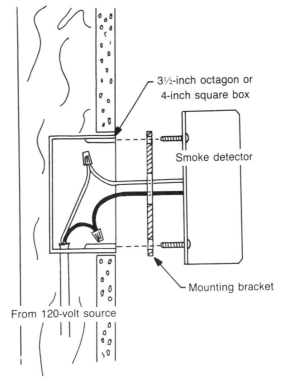

Fig. 7-2. Smoke detector wired into house circuit.

DIMMER SWITCHES

Living and dining areas often have overhead lights that are too bright for all situations. Dimmer switches allow you to vary the intensity to suit the mood of the occasion (Fig. 7-4). Note that most

Fig. 7-3. Back view of smoke detector showing battery.

dimmer switches can be used only on incandescent lamps—the conventional light bulbs normally found in our homes. Special dimmer switches are required for fluorescent lights.

Normally these switches allow you to adjust the light from bright, to medium dim, or anything in between. They have the additional advantage of reducing electrical consumption when turned to any point below full bright.

Dimmer switches are easy to install. First, turn off the breaker for that circuit. Remove the existing switch and connect the wires to the dimmer switch. Mount the switch and control knob, and the installation is complete.

THREE-WAY SWITCHES

In the past, darkened stairways have been the cause of painful injuries. Lights in hallways become much more useful if they can be switched on and off from either end of the hall. Sometimes it would be handier if a garage light could be turned on and off from inside the home, as well as the garage. Three-way switches can

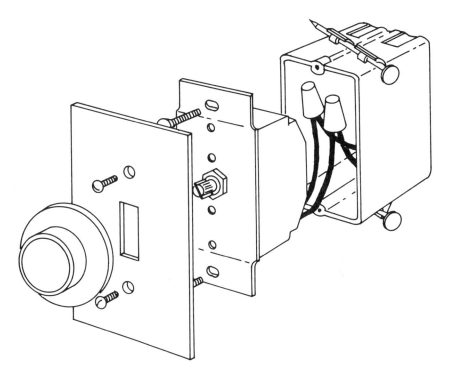

Fig. 7-4. Dimmer switch used to vary light intensity. (COURTESY OF SQUARE D COMPANY)

help in these situations, and they offer some inexpensive insurance against stumbling over some obstacle in the dark.

Despite their name, these switches operate in pairs to control a light or receptacle from two different places, not three. Three-way switches have three terminals; two terminals of the same color, usually brass or silver colored, and one of another darker color, often black (Fig. 7-5). There is no top or bottom or ON-OFF marking. The important thing is to locate the odd-colored terminal. This darker terminal is often referred to as the *common terminal*.

Connect the hot wire from the source to the common terminal of one switch, and the common terminal of the other switch to the hot wire going to the light. Now you have two unused terminals of the same color on each switch. Just run two hot wires from the two terminals on the first switch to the remaining two terminals of the second switch. Which wire goes to which terminal doesn't matter. Notice in Figs. 7-6 and 7-7 that the ground wire is not shown. If you are using plastic boxes, just crimp the bare ground wires together.

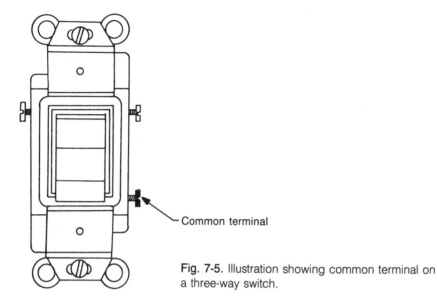

Fig. 7-5. Illustration showing common terminal on a three-way switch.

If a light needs to be controlled from three or more locations, use a combination of two three-way switches and one four-way switch (Fig. 7-8). For each additional location, just add another four-way switch. Four-way switches have four terminals and no ON-OFF markings. They should be installed in the middle of the run between the two three-way switches.

TIMERS

Lights and other devices can be set to go on and off automatically, day or night, by switches called *timers*. The wiring on 120-volt timers is exactly like that for any other switch. Connect the neutral (white) wires together, and the two hot (black) wires to the switch terminal.

Hot-water heaters consume a large portion of the energy used in homes. The installation of a timer in this circuit can reduce an electric bill up to 30 percent. Be sure to use a timer with a high enough current rating. Make certain the power is off and locked, if possible. Then check with a voltage tester to be sure you turned off the right circuit before you begin any work.

Timers are just automatic switches and can be mounted on the wall near the water heater or alongside the breaker box. A 115-volt water heater will be fed by one hot (black) wire, one neutral (white) wire, and one ground (bare) wire. Wire it as described previously. However, a 230 volt water heater will be supplied by

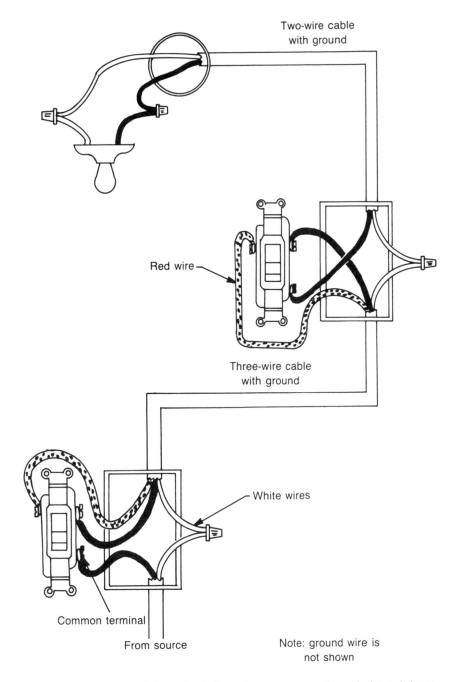

Two-wire cable
with ground

Red wire

Three-wire cable
with ground

White wires

Common terminal

From source

Note: ground wire is
not shown

Fig. 7-6. Three-way switches wired where the power goes through the switches to a light.

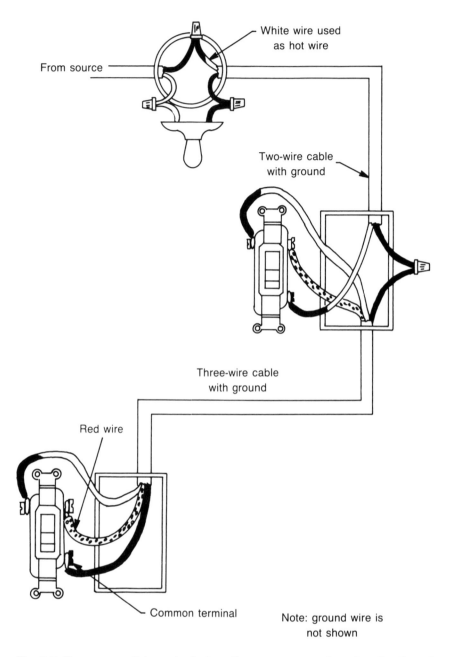

White wire used
as hot wire

From source

Two-wire cable
with ground

Three-wire cable
with ground

Red wire

Common terminal

Note: ground wire is
not shown

Fig. 7-7. Three-way switches wired where the power comes from the direction of the light.

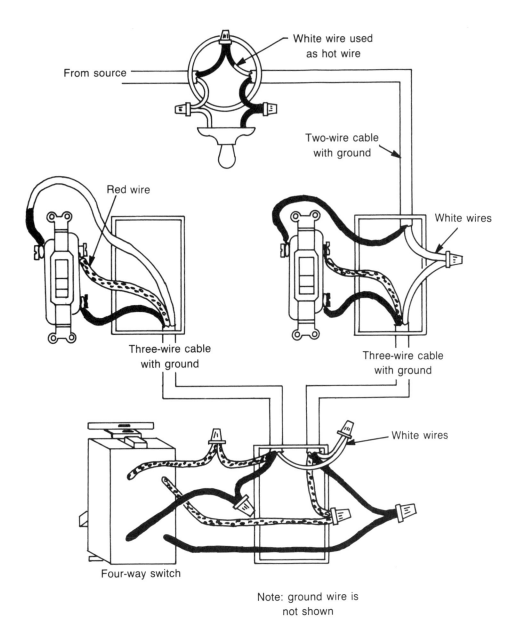

From source

White wire used
as hot wire

Two-wire cable
with ground

Red wire

White wires

Three-wire cable
with ground

Three-wire cable
with ground

White wires

Four-way switch

Note: ground wire is
not shown

Fig. 7-8. Four-way switch wired between two three-way switches.

two hot, normally black or red, wires and a bare ground wire. There
probably will be no white wire. Terminal markings inside the timer
might read "line 1," "load 1," "line 2,and " "load 2." The water
heater will be the load, so connect the two hot wires coming from
the heater to the two terminals marked "load." "Line" means supply
voltage, so correct the two hot wires coming from the breaker to

Fig. 7-9. Model TC-2 Timer Control. Terminals located under insulated cover.
(COURTESY OF PARAGON ELECTRIC COMPANY, INC.)

the terminals marked "line." Connect the bare ground wire to a grounding screw (often green) found in the cabinet of the timer. Other timers such as the one shown in Figs. 7-9 may have terminals marked with numbers 1 through 4. In this case, the timer is wired as shown in Fig. 7-10.

Manufacturers always provide instructions for their particular timers, so it is important to follow their guidelines. After you have installed the timer, set the trippers for the times desired, then turn the clock face clockwise until the current time is indicated by the black pointer (Fig. 7-11). Return the insulated cover to its place, then turn the power back on.

INSTALLING GROUND-FAULT INTERRUPTERS

The grounding wire in three-prong receptacles does lower the risk of shocks when used with a cord with a ground. Under normal conditions, when the receptacle is being used, the amount of current going to and returning from the receptacle is identical. If the amount

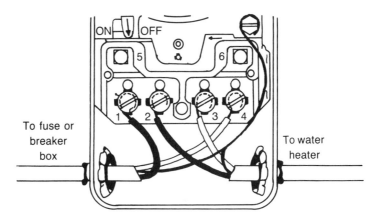

Fig. 7-10. Wiring diagram. (COURTESY OF PARAGON ELECTRI COMPANY, INC.)

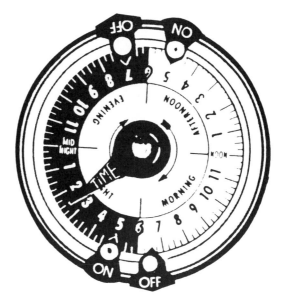

Fig. 7-11. Typical TC-2 program. ON: 5:00 a.m., OFF: 6:30 a.m. ON: 5:00 p.m., OFF: 7:00 p.m. (COURTESY OF PARAGON ELECTRIC COMPANY, INC.)

of current leaving the receptacle is not absolutely the same amount of current going in, the tool, appliance, or wiring is defective, allowing some of the current to leak to ground. This is called a *ground fault*. It could be particularly dangerous around damp locations such as bathrooms, or garages, or outdoors. An inexpensive device that offers protection from such circumstances is called the ground-fault circuit interrupter (GFCI), shown in Fig. 7-12. The GFCI constantly monitors the amount of current to and from the receptacle. If there

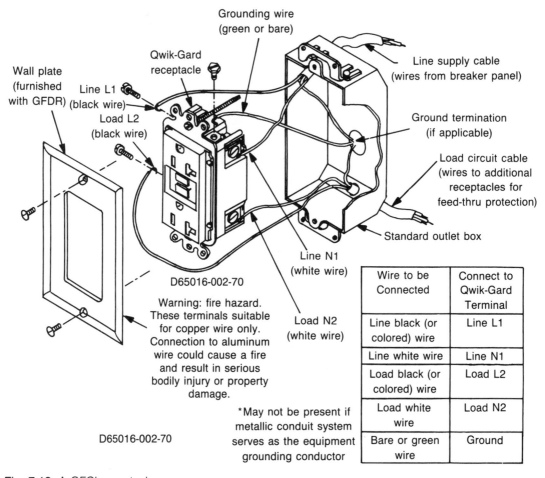

Wire to be Connected	Connect to Qwik-Gard Terminal
Line black (or colored) wire	Line L1
Line white wire	Line N1
Load black (or colored) wire	Load L2
Load white wire	Load N2
Bare or green wire	Ground

Wall plate (furnished with GFDR)

Line L1 (black wire)

Load L2 (black wire)

Qwik-Gard receptacle

Grounding wire (green or bare)

Line supply cable (wires from breaker panel)

Ground termination (if applicable)

Load circuit cable (wires to additional receptacles for feed-thru protection)

Standard outlet box

Line N1 (white wire)

Load N2 (white wire)

D65016-002-70

Warning: fire hazard. These terminals suitable for copper wire only. Connection to aluminum wire could cause a fire and result in serious bodily injury or property damage.

D65016-002-70

*May not be present if metallic conduit system serves as the equipment grounding conductor

Fig. 7-12. A GFCI receptacle. (COURTESY OF SQUARE D COMPANY)

is the slightest deviation, about five thousandths amp, it instantly opens the circuit, shutting off the power from the receptacle in about 1/40 second.

Install ground-fault receptacles like an ordinary receptacle, except that they come prewired. Just connect the leads marked "line" to the wires coming from the power source: black to black, white to white, and green to ground. If you will wire in more receptacles downstream, use the leads marked "load" to make the connection. If no further connections are necessary, install insulated wire nuts on each lead marked "load" and fold out of the way.

The GFCI receptacle and any receptacles wired downstream are the only ones with ground-fault protection. Receptacles connected upstream, toward the power source, will not be protected.

After the installation is complete, you probably will have to push the reset button to activate the outlet. If the unit does not reset, turn the power back off and check for nicked wires and proper installations of receptacles downstream. Unplug any appliances from outlets downstream. If the unit does reset, then plug a lamp into the receptacle and press the test button. The button should pop out and the lamp go off. If it doesn't, the ground fault receptacle is probably bad. If the button pops out but the light doesn't go out, the unit is wired incorrectly. Check the "line-load" connections on the back.

Circuit-breaker GFCIs can be installed in the breaker box (Figs. 7-13 and 7-14). They perform the same way as other GFCIs shutting off the entire circuit and providing protection for a number of outlets. Turn the power off when installing either GFCI. If you are installing the circuit breaker type, the entire panel must be dead, so turn off the main breaker. The code requires GFCI protection on all 15- and 20-amp receptacles installed for general use in bathrooms, garages, and outdoor areas. Install the circuit breaker GFCI just like a conventional breaker, except that it has a grounding wire that you must connect to the grounded neutral busbar in the panel. Because this unit is installed at the power source, it will provide ground-fault protection to all of the receptacles on the circuit. The code requires GFCI protection in new homes, but homeowners with older dwellings also should consider this inexpensive safety device as a necessary addition.

CONNECTING TO CIRCUIT BREAKERS

Most overcurrent protection today is provided by circuit breakers. This method is preferred over fuses because fuses must be replaced but circuit breakers can simply be reset to restore power.

To convert an old fused system to circuit breakers requires the replacement of the panel, either a subpanel supplying power to branch circuits, or the main service-entrance panel, or both. If the service-entrance panel has enough capacity to handle the new subpanel, then you will need to replace the subpanel only. If not, then you should replace both panels. When locating a new service entrance panel, mount the new panel as close to the meter as possible. The best location might be directly below the meter on an outside wall or on the inside wall behind the meter.

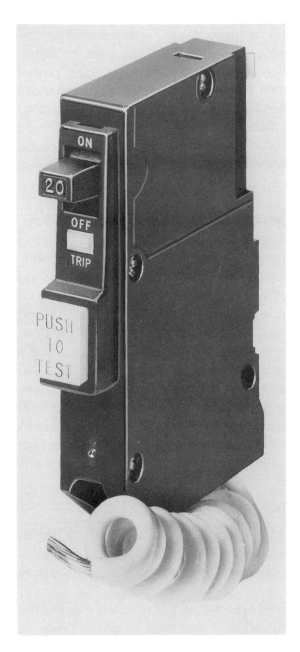

Fig. 7-14. A GFCI circuit breaker installed to protect a branch circuit. (COURTESY OF SQUARE D COMPANY)

Fig. 7-13. A GFCI circuit breaker. (COURTESY OF SQUARE D COMPANY)

Panels come in different sizes (Figs. 7-15 and 7-16), and models are available with flush and surface mounting the flush mounting is designed to fit in between studs so that the outer surface is flush with the finished wall. If there is enough room, install the new panel where the old one is.

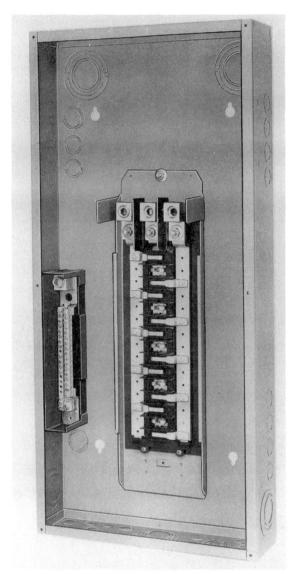

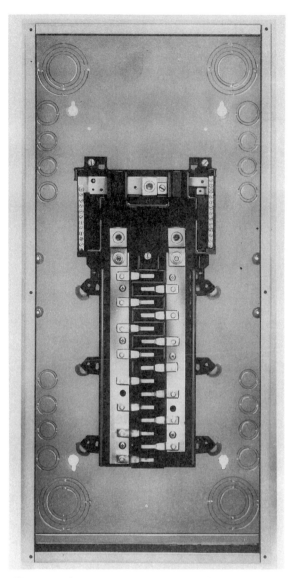

Fig. 7-15. A cabinet showing neutral assembly located off to one side. (COURTESY OF SQUARE D COMPANY)

Fig. 7-16. A cabinet without breakers showing neutral assembly positioned on both sides of where the main breaker will be. (COURTESY OF SQUARE D COMPANY)

If you replace the service entrance panel, the power will have to be turned off by the utility company. Plan to be without electricity for a day or so. After the electricity has been turned off, disconnect the wires coming into the old panel. Label the wires so you will be able to identify each circuit, and mark them in the new panel. Tie the wires up out of the way so you will have room to work. Remove the old panel and hold the new one in place.

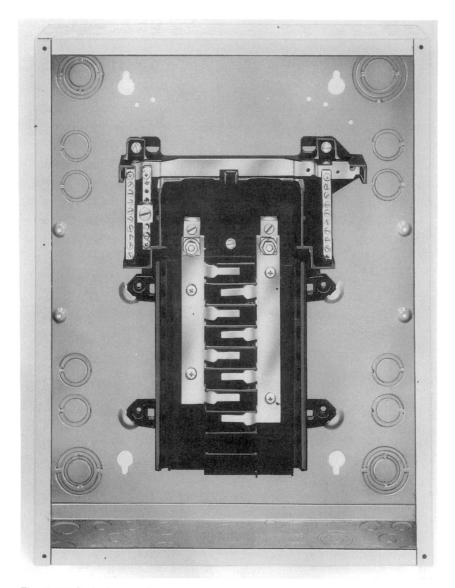

Fig. 7-17. An indoor enclosure. (COURTESY OF SQUARE D COMPANY)

The new panel will probably be empty (Fig. 7-17). It might not even have the main breaker (Figs. 7-18 and 7-19). Purchase the individual circuit breakers separately, selecting them according to the needs of each circuit.

Mark the holes for mounting the panel. If you are mounting the panel on a concrete block wall, use anchor bolts. A ¼–20 size should work. Align the new panel so it will be straight and level.

Fig. 7-18. 150-225 amp and 70-125 amp main breakers. (COURTESY OF SQUARE D COMPANY)

Make the installation neat. After the panel is securely mounted, install the circuit breakers (Fig. 7-20) and connect the wires. Check over the installation. Make sure wires are secure and tight in their connections. Install the cover (Figs. 7-21 and 7-22) and label each circuit in the place provided. You should now be able to restore the power.

If you cannot mount the new panel in place of the old one, then you will need a junction box. A box about 8 × 10 inches is a good size. You might be able to use the old panel as a junction box. Remove the old incoming service wires, and disconnect all of the individual circuit wires. If the old panel is equipped with a cover or door where the inside can be closed off, then remove and discard everything inside—fuse holders, neutral busbar, terminals, etc. Install knockout seals in any unused holes.

If the old panel will not work as a junction box, simply remove the old panel completely and install an 8- x -8 inch or larger junction box. Then run the old wires into this junction box. Next run wires from the new panel to the junction box, one cable for each circuit and the same size as the old wires. Make the connections inside the junction box using wire nuts, connecting black wires to black

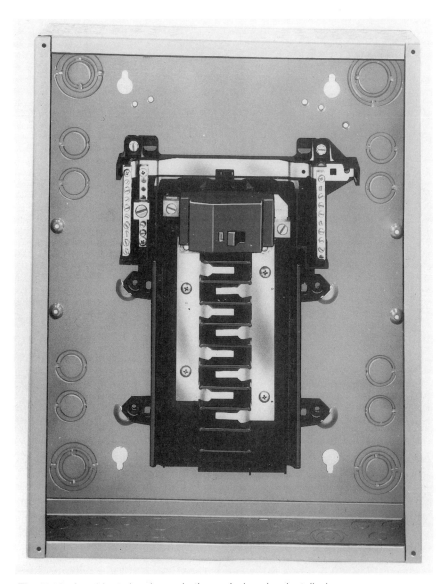

Fig. 7-19. A cabinet showing only the main breaker installed. (COURTESY OF SQUARE D COMPANY)

wires and white wires to white wires. The junction box will not have a grounding busbar, but the box itself must be grounded. Connect the wires from the junction box in the new service panel to the individual breakers. Check all connections, replace the covers, and restore the power.

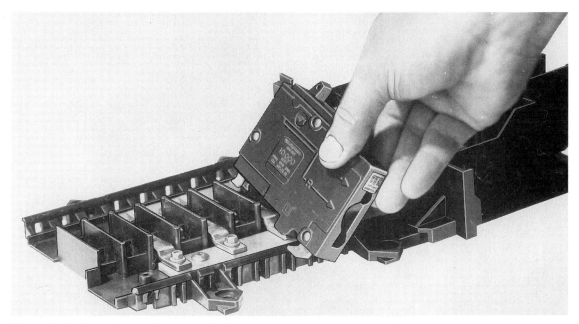

Fig. 7-20. A circuit breaker being installed. (COURTESY OF SQUARE D COMPANY)

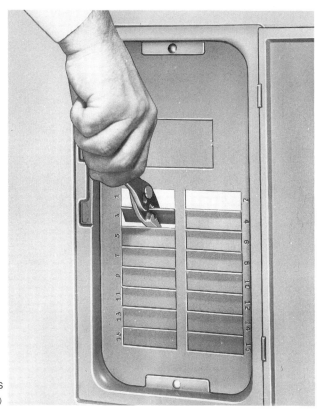

Fig. 7-21. Shutter-type twistouts provide spaces for circuit breakers. (COURTESY OF SQUARE D COMPANY)

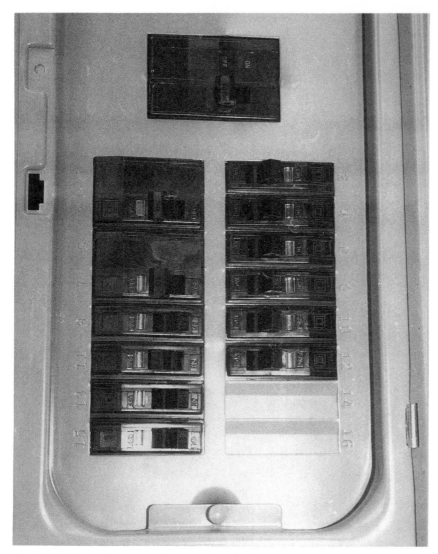

Fig. 7-22. Circuit breakers with cover installed.

8

Automatic
Garage Doors

A NUMBER OF GARAGE DOOR OPENERS ARE ON THE MARKET, AND IN-
stalling one yourself can be very satisfying. At first thought, it might
sound like a sizable undertaking, but the average homeowner can
install one in a little over a half a day with only a basic set of tools.
A good-quality door opener can be purchased for about $150 to
$200 (Fig. 8-1). All systems come with a small remote transmitter
that sends a signal to the receiver in the unit inside the garage.
If more than one car will use the garage, additional transmitters
are available for about $30 each.

Models vary, depending on whether the installation will be for
a track or trackless garage door (Fig. 8-2). The unit should have
a light that comes on when the door begins to open. These are
usually timed to go off a few minutes after the door is closed. Safety
features normally include some type of automatic reversing mechanism if the door fails to close properly because of some obstruction,
such as the back of the car or a child's tricycle.

Most units are chain driven; however, one company has a
plastic track, or tape, that works very well. The drive motor can vary
from $\frac{1}{4}$ to $\frac{1}{2}$ horsepower, depending on the size of the door.

Some models, such as the Stanley Premier model (Fig. 8-3),
have added features. This includes a vacation switch, which turns

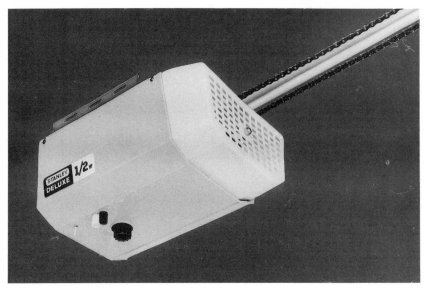

Fig. 8-1. Stanley's Deluxe Model ½-hp opener comes equipped with all of the basic features. (COURTESY OF STANLEY AUTOMATIC OPENERS)

off the power, disabling the opener. This model also comes with a two-button transmitter equipped with Stanley's Signal-Block. The extra button allows you to shut off the door opener from your car as you drive away. This button offers additional security, as well as prevents stray radio signals, such as aircraft transmitters, from activating the door opener. A work light is another feature. The opener is equipped with a pull switch that allows the opener light to be used as a work light for the garage area.

A typical installation for a garage door opener would have a 120-volt receptacle mounted near the door operator. Sometimes they are connected to the light circuit. In this case, a pull chain may be necessary to operate the light because the switch will have to be on all the time. If this is not practical, you can tap into the circuit upstream from the switch where you will have full-time power. Run #12/2 cable over to a suitable mounting surface near the operating unit, mount the box, and install the receptacle. This way the door operation will not interfere with the normal garage lighting.

While some openers are designed more for the professional installer, some companies offer their product for the do-it-yourself homeowner. They provide ample instructions and liberal illustrations to enable almost anyone to install his own opener. Stanley even

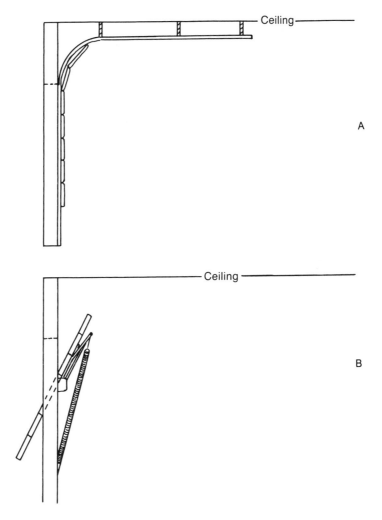

Fig. 8-2. Two types of garage doors: track (A) and trackless (B). Be sure the model opener purchased is for the type of door it is to operate.

provides a toll-free hotline if you run into problems with the installation.

TYPICAL PARTS AND ASSEMBLY

It is important to have some idea of what makes up a garage door opener before you began the assembly (Fig. 8-4). Remember, garage doors are extremely heavy, even though they are usually lifted with only a little effort. The easy manual operation is possible through powerful balancing springs or torsion springs. Do not

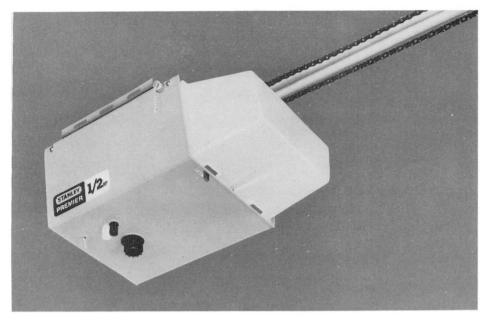

Fig. 8-3. Stanley's Premier Model features extras such as a vacation switch, work light, and a two-button transmitter with Signal-Block. (COURTESY OF STANLEY AUTOMATIC OPENERS)

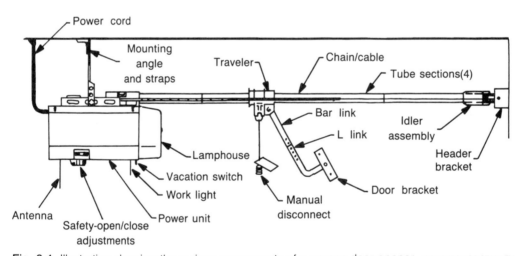

Fig. 8-4. Illustration showing the various components of a garage-door opener. (COURTESY OF STANLEY AUTOMATIC OPENERS)

disconnect or alter these springs; they are required for the safe operation of the door.

It is a good idea to remove any ropes or cords attached to the garage door. They could entangle someone later and cause an injury. Rings, watches, and loose clothing also pose a safety hazard when installing or servicing garage doors. Always follow the owner's manual closely and work carefully and deliberately, always keeping safety in mind.

Garage door openers have instruction manuals that are usually easy to follow. Basically, the first step of the project would be to assemble the unit. Position the drive chain around a sprocket, and attach tracks or rails to the control unit.

Locate the center of the garage door. Try to be precise. There is only one point of lift, so the load should be balanced. Next raise the door far enough to establish the highest point of travel and mount a support bracket, usually 2½ inches above the highest point of door travel (Fig. 8-5). Mount the drive unit's brackets, then attach the assembled unit to its brackets (Fig. 8-6). Adjust door travel by setting limit switches or control knobs, depending on the model controller installed.

ADDING LIGHTS

For added convenience, you could install a light outside that would light the driveway (Fig. 8-7). You could wire the light into the timed light on the control unit, and the light would go on when the door is activated and off when the inside light times out.

A handy addition to this same light circuit would be a pilot light mounted inside the home that would indicate when the door is open (Fig. 8-8). Run the cable to a box the same as if it was feeding a receptacle. You can use a socket with a small-wattage bulb, or you can install a duplex receptacle and use a plug-in night light as the pilot light.

Be sure to unplug the door opener from its receptacle; you don't want to work on a live circuit. Next remove the cover from the control unit, and locate the wires going to the light and a suitable knockout for the cable coming in. Make sure you have a clear route, and then feed the wire to the pilot light box. Then mount the box and run the cable back to the control unit or handy box going to the control unit.

If you are tapping into the outside light, feed the cable through the knockout hole using cable clamps, cut the wires to the light, and make the connections. Using a wire nut, connect the black wires to black wires and the white wires to white wires. You should have

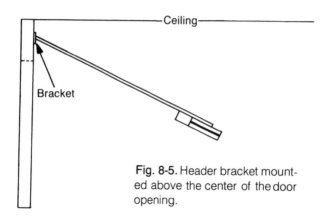

Fig. 8-5. Header bracket mounted above the center of the door opening.

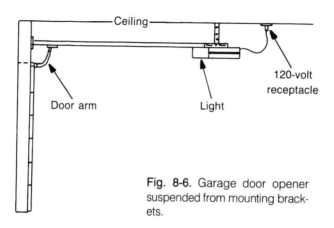

Fig. 8-6. Garage door opener suspended from mounting brackets.

three of each color. Plug in the door controller, and replace the cover. The light and/or pilot light should come on if the door is open.

If a low-voltage pilot light is more appealing, you can wire a 12-volt transformer into the wires going to the light and use a 12-volt pilot light instead. If you use bell wire to supply the power to the light, no box would be necessary. Simply drill a ½-inch hole in the wall, and pull about 6 or 8 inches of bell wire through to make the connections. Drill a hole in a blank electrical cover to mount the light. Use a very small pilot drill for the mounting screws. After the pilot light is mounted, connect the bell wire to the 12-volt side of the transformer. Use #18 lamp cord to make the splice between the wires going to the light in the door operator and the 120-volt side of the transformer.

If the door opener light is not suitable, you can use a 12-volt transformer, a 12-volt pilot light, and a magnetic switch of the kind

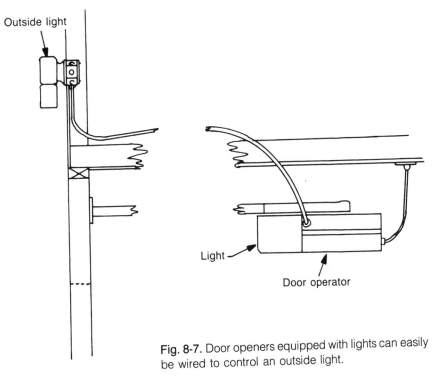

Fig. 8-7. Door openers equipped with lights can easily be wired to control an outside light.

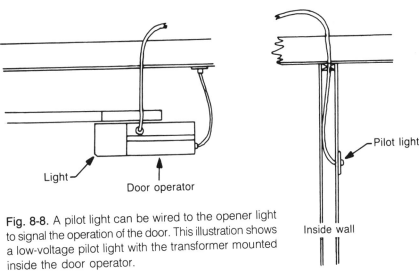

Fig. 8-8. A pilot light can be wired to the opener light to signal the operation of the door. This illustration shows a low-voltage pilot light with the transformer mounted inside the door operator.

used in burglar alarms (Fig. 8-9). You'll need a length of #18 lamp cord with a plug to supply the 120 volts to the high side of the transformer. The wire going to the pilot light and magnetic switch, or switches, for two or more doors, can be bell wire or #24 speaker

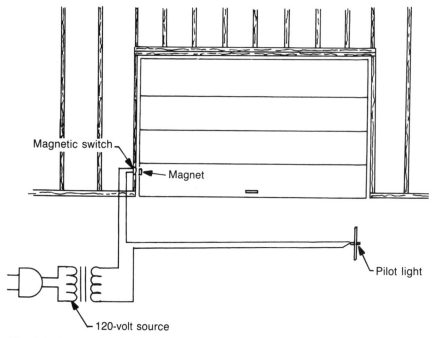

Fig. 8-9. A magnetic switch used to activate a low-voltage pilot light.

wire. Stranded wire is better if a number of bends or twists is necessary.

If you want the pilot light to go on when the door is open, use a *normally open* magnetic switch. This way the light will be off most of the time and will be easily noticed when the door is open. The only problem is if the bulb fails and is not discovered; then the door could be left open unknowingly.

A *normally closed* switch will keep the pilot light on all the time and then go out when the door is opened. The advantage of this arrangement is that when the light is out the door will either be open or the bulb will be burned out. However, any indicator or warning light that is on when it is normally off seems to attract more attention than the other way around. Just check the system from time to time to make sure it's working.

Probably the first step in installing this system is to find a location for the pilot light. Because it will be low voltage, no box is necessary, but it should provide easy access for the wire. Then decide on the location for the transformer. It should be near a 120-volt receptacle and out of normal reach. After you have determined these locations, drill a ½-inch hole in the wall for the back of the pilot light, and run the bell wire or speaker wire from the transformer location to the hole. Leave a few inches sticking through the hole to make the connections to the light.

Drill a hole in a blank electrical cover and mount the light. The leads on the bulb will probably be very short, so you will be able to make the connections easier if you solder longer wires to the leads prior to making the connections with wire nuts. To mount the light assembly, feed the wires back in the hole along with the back of the light, align the plate, and drill small pilot holes for the mounting screws. Install the screws, but don't overtighten them if they're going into wallboard or plaster.

Mount the magnet on the door and the magnetic switch on the door frame. The switch must be within ¼ inch of the magnet in order to operate. The door should be completely closed. Next mount the transformer and connect the bell wire through the switches to

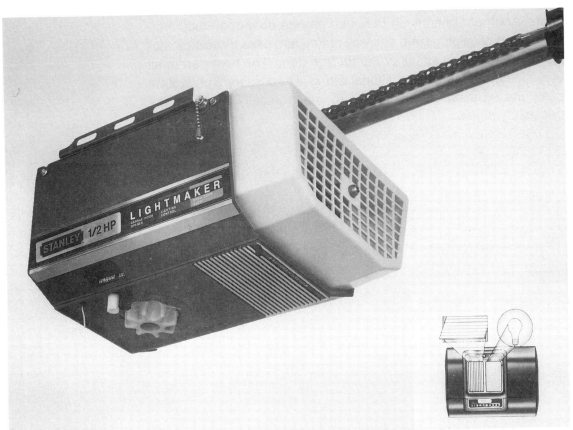

Fig. 8-10. Stanley's Light Maker™ Model provides options that include controlling outside and inside lights using the existing house wiring. (COURTESY OF STANLEY AUTOMATIC OPENERS)

the low (12-volt) side of the transformer. Connect the #18 lamp cord to the 120-volt side of the transformer and install a plug on the other end. Plug in the transformer and make any magnet-to-switch adjustments necessary.

While you're running the wire, you might consider installing a burglar alarm. Just connect a 12-volt buzzer or horn and an on and off switch in parallel with the pilot light. When the switch is off, the system will operate normally, but when the switch is on, the horn will blow when the door is opened and the pilot light comes on.

If your home is not equipped with a garage door opener and you want all of the extra features, you might want to buy an opener that offers these features built in. Stanley U-Install Garage Door Openers introduced the Light Maker (Fig. 8-10) in July 1987. This opener has all of the basic features plus the capability for controlling a plug-in lamp, a switch that turns on a light such as an outdoor or front door light, and a three-way wall switch that controls hallway lights. Another option is an in-house garage-door-open indicator. This little device plugs into any wall outlet, and uses three indicator lights, to signal the status of the garage door. The best part of all these features is that no additional wiring is necessary. The system uses the existing house wiring to electronically transmit the necessary signals.

9

Adding a
Shop Circuit

A BRANCH CIRCUIT FEEDING A SHOP CAN BE APPROACHED THE SAME as any other circuit. If the shop is not attached to the house, consider underground service. Determine how much power will be needed for motors, saws, drills, electric heater, air conditioners, etc. Should it be 120 or 240 volts? If current requirements are large, you can run heavier wires from a subpanel main breaker in the service-entrance panel to a subpanel in the shop (Fig. 9-1). This panel would contain the breakers for the individual circuits in the shop (Fig. 9-2). Subpanels have the advantage of providing convenient access to the circuit breakers and eliminating the need for a number of long circuit runs.

If underground wiring is impractical and you must run it overhead from the house to the shop, then you must follow the code's restrictions (Fig. 9-3). The wires must be kept at least 10 feet above the finished grade, sidewalk, or any deck area that is accessible to foot traffic. If the wires travel over a residential driveway, they must be suspended at least 12 feet above the drive. The code further states that no trees or other vegetation may be used to support overhead conductors or equipment.

Wires running from the subpanel are connected to the breaker in the service entrance panel just like any 120/240 branch circuit.

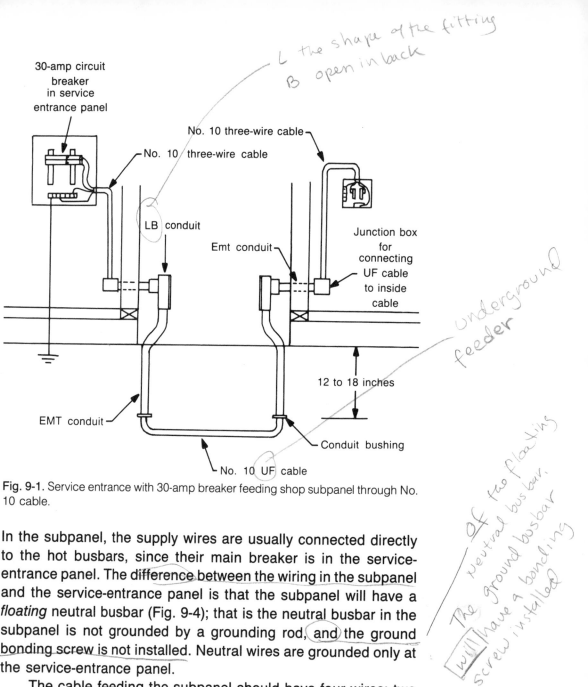

Handwritten annotations:
- the shape of the fitting B open in back
- underground feeder
- Of two floating neutral busbar, The ground busbar will have a bonding screw installed

30-amp circuit breaker in service entrance panel

No. 10 three-wire cable

No. 10 three-wire cable

LB conduit

Emt conduit

Junction box for connecting UF cable to inside cable

12 to 18 inches

EMT conduit

Conduit bushing

No. 10 UF cable

Fig. 9-1. Service entrance with 30-amp breaker feeding shop subpanel through No. 10 cable.

In the subpanel, the supply wires are usually connected directly to the hot busbars, since their main breaker is in the service-entrance panel. The difference between the wiring in the subpanel and the service-entrance panel is that the subpanel will have a *floating* neutral busbar (Fig. 9-4); that is the neutral busbar in the subpanel is not grounded by a grounding rod, and the ground bonding screw is not installed. Neutral wires are grounded only at the service-entrance panel.

The cable feeding the subpanel should have four wires: two hot wires, one neutral, and one ground wire. Make circuit connections in the subpanel by connecting first the hot wires to the breaker, then the white wire to the floating neutral busbar, and finally the ground wire to the grounding busbar, which is bonded to the panel. This way the panel is grounded while the neutral is isolated from the panel.

Fig. 9-2. Shop supanel. (COURTESY OF SQUARE D COMPANY)

Be sure to use the proper cable. If the service is to be underground, you can use type UF (underground feeder) cable if you protect it by breakers in the service-entrance panel. You can bury the cable directly in the ground. For residential installations, the minimum depth is 12 inches, provided that the run is protected

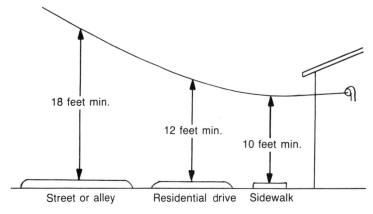

Fig. 9-3. Minimum clearance for overhead wires.

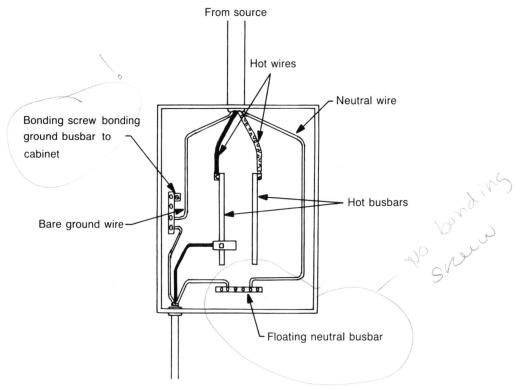

Fig. 9-4. Subpanel with floating neutral busbar.

by a breaker of 30 amps or smaller, otherwise, the minimum depth is 24 inches. Where the cable enters a building, an LB is a convenient fitting to use. The *L* describes the shape of the fitting while the *B* means it opens in the back.

When you are running the cable, don't pull it tight, but leave plenty of slack in the trench. Don't backfill the trench with sharp rocks. Surround the cable with a few inches of sand, and if there is any chance of striking it with a shovel, bury a running board for protection.

If the service is to be overhead, use weatherproof wire. This wire cannot be used for normal wiring and it can only be used overhead outdoors. Overhead outdoor wiring must not only be big enough to carry the current for the circuits, it also must be large enough to support its own weight even if coated with ice. The code requires a minimum of #10 for spans of 50 feet or less, but for colder areas, a size larger (#8) is more realistic.

Wires contract in cold weather, so allow a little slack for overhead wires run in warm weather. If you do not, supporting insulators might be pulled from buildings when the temperature drops.

Use the proper weatherproof heads for leaving and entering the buildings (Fig. 9-5), and maintain the minimum clearances as described by the code.

After you have mounted the subpanel, run the individual circuits to it just like for any other room, except that you will want to have

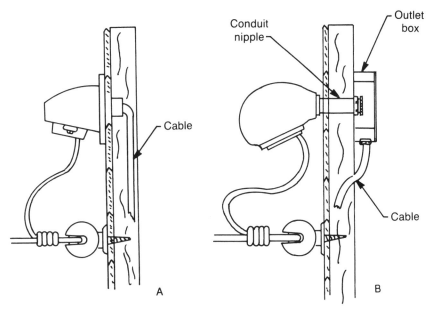

Fig. 9-5. Two different methods of bringing wires into a building. (A) Simple entrance cap requiring only one hole. (B) Entrance cap with nipple entering an outlet box.

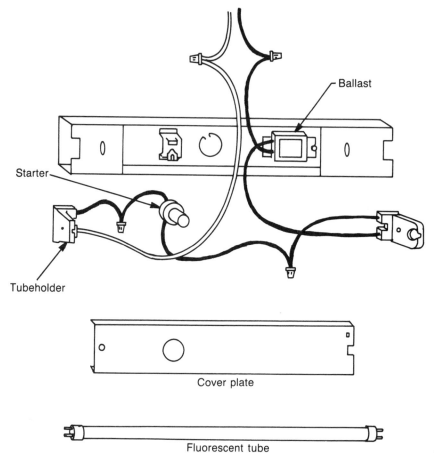

Ballast

Starter

Tubeholder

Cover plate

Fluorescent tube

Fig. 9-6. Breakdown of a fluorescent light fixture.

extra outlets along the workbench. If two or three motors will be plugged in and running at the same time, it might be better to split the number of outlets into two circuits. Have an outlet for plugging in a floating arm work light for the work counter. Some are equipped with a magnifying lens for close work.

A fluorescent ceiling fixture can provide economical lighting for the general area. Fluorescent lights use much less energy than ordinary incandescent bulbs, and they last about four times as long. They are available in both decorative and utility styles. The utility model is just a metal box with a number of knockouts for different entries for the cable, the ballast (transformer), and the ceramic mounts for the tubes (Fig. 9-6). All of the parts can be purchased separately for replacement later.

You can use screws to attach these fixtures to ceiling joists, or you can mount them on an outlet box just like an ordinary light fixture. You can use a pull-chain switch, or wall switch to control them. Most dimmer switches will not work on fluorescent lights.

10

Adding
Telephone Jacks

MORE AND MORE, HOMEOWNERS ARE INSTALLING THEIR OWN TELE-
phone systems because of the high costs charged by telephone
companies and the ease and simplicity of the installation itself. Rates
charged by the phone company can vary from $50 to $75 an hour
after an initial service charge of approximately the same amount.
The materials are inexpensive, so almost all of the costs are for
labor. Anyone who can use a screwdriver and twist a wire around
a screw can add or upgrade an existing telephone service.

TELEPHONE SERVICES AND VOLTAGES

Voltages powering a residential telephone service may vary
from 48 volts dc to as much as 105 volts ac. Normally these voltages
are not dangerous, but you should take some precautions. If you
have a pacemaker, you should not work on your telephone system
at all. Avoid working on a live telephone line. Take the telephone
off the hook or disconnect it at the main terminal box. Never work
on any wiring during a thunderstorm, and if holes are to be drilled
in walls, be alert for electrical wires and water pipes.

Residential telephone services enter the house in one of two
ways. The first way is through an overhead service connected to

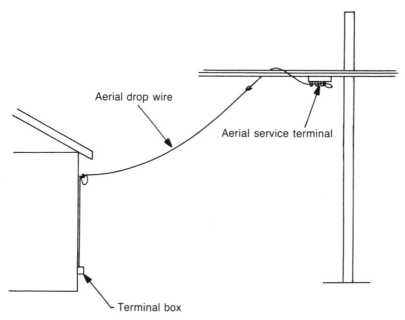

Fig. 10-1. Aerial telephone service.

a nearby telephone pole (Fig. 10-1). This method usually consists of one cable containing only two wires. The pair of wires provides service for one telephone number. If an additional number is desired, usually another cable must be run from the pole to the house. If the local network is unable to provide enough cable pairs, they may install line-splitting equipment. So one cable pair can provide service for two different telephone numbers.

The other method of service is underground (Figs. 10-2 and 10-3). Because of the problems involved in digging it back up, an underground service usually consists of two cables containing three pairs of wires in each cable. With six pairs of wires coming in, up to six separate circuits (telephone numbers) can be installed. You probably won't have six different telephone numbers for your home, so if one pair of wires happens to go bad, there should be enough unused pairs to quickly restore the service.

Whether the service is overhead or buried, the pair of incoming wires will terminate in a plastic or metal box installed by the telephone company (Fig. 10-4). This terminal box may be located on an inside or outside wall of the residence, and unless other arrangements are made, this is as far as the telephone company goes.

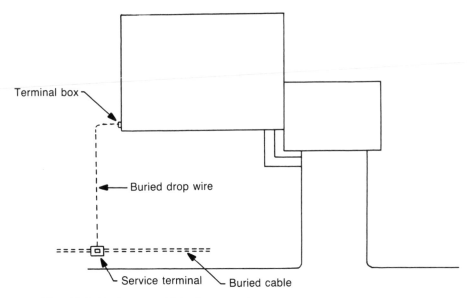

Terminal box

Buried drop wire

Service terminal

Buried cable

Fig. 10-2. Underground telephone service.

Nearly all of the terminal boxes are equipped with fuses (Fig. 10-5). These fuses protect your telephone and associated wiring from damage caused by high-voltage surges, such as those generated by lightning strikes. Older installations might have boxes not equipped with fuses. In these systems, the wires going to the telephone include a ground wire. The surge of electricity could travel through the telephone before it would dissipate through the ground conductor to earth ground. Often the telephone company will replace a fuseless box without charge.

CIRCUIT CONNECTIONS

From the terminal box, run the wires to the wall jacks mounted at the desired locations wall jacks are sometimes called PMCs (plug-in modular connectors). The number of telephones you install will depend on your own personal needs, but the number that will ring depends on the amount of power provided by the telephone company. This is called the *ringer equivalence number*. Normally there is enough power to ring about five phones (5 REN). On the underside of the telephone, there should be a label indicating the ringer equivalence (Fig. 10-6). Just add up all of the REN values of your phones, and as long as the total doesn't exceed 5, they should all ring. If the value is greater than 5, they should still work, but not all of them may ring.

Fig. 10-3. Marker showing buried cable.   **Fig. 10-4.** Weatherproof terminal box.

Wires supplying service to the individual jacks may be connected directly to the terminal box, where the service enters the house (see Fig. 10-7), or they may connect to an existing phone jack (Fig. 10-8). Choose the location that is most convenient. You may need only two wires now, but for future use it is a good idea to run extra wires. A typical number would be two pairs of wires (Fig. 10-9). Most homes constructed in the last few years have had the telephone service installed during the framing stage (Fig. 10-10). The wires are run inside the walls, making the installation much more attractive.

Another option is to run the wires in a crawl space under the floor (Fig. 10-11). Then bring them up through the floor close to the baseboard of an inside wall. You also can choose to run the wire along an exterior wall to the desired location (Fig. 10-12) or around or beneath a roof overhang (Fig. 10-13). Inside the house, you can conceal the wires behind moldings and under carpets along baseboards, or simply staple them neatly along the tops of

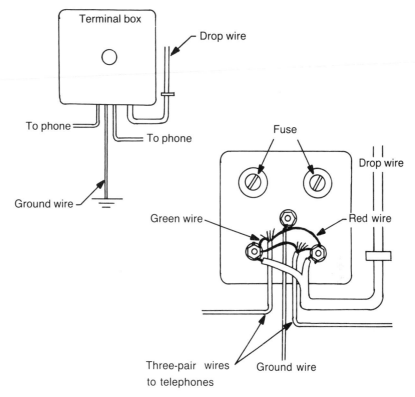

Fig. 10-5. Terminal box with typical wire connections to two telephones.

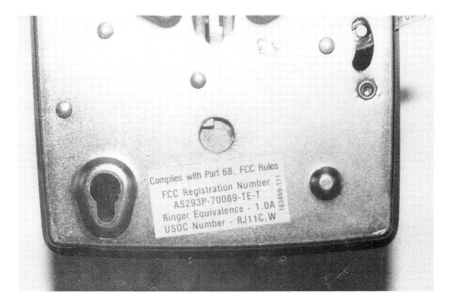

Fig. 10-6. Label on the back of the phone gives the REN.

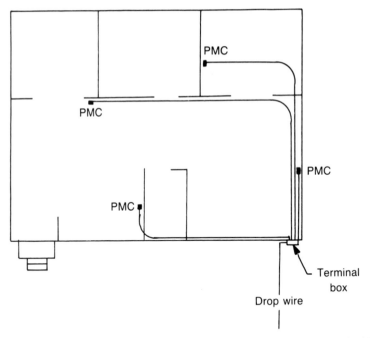

Fig. 10-7. Phone jacks wired where each jack is connected directly to the terminal box.

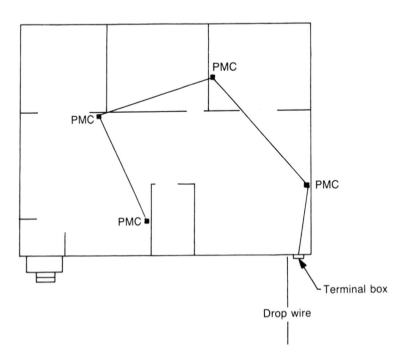

Fig. 10-8. Phone jacks wired in series.

148

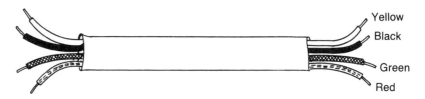

Fig. 10-9. A popular inside telephone wire consisting of two pairs.

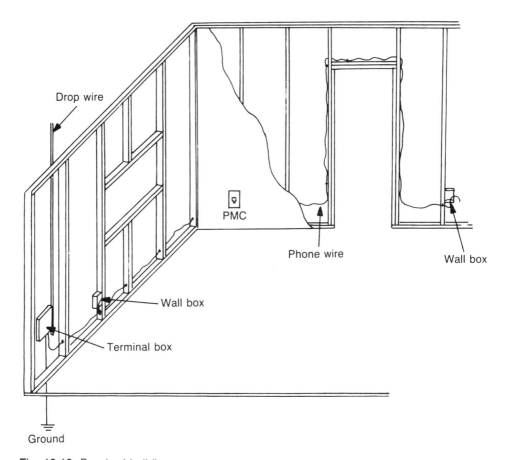

Fig. 10-10. Prewired building.

baseboards and doorways (Fig. 10-14). You might find it convenient to run the wire through a wall to the next room (Fig. 10-15). Just try to make the wiring as inconspicuous as possible.

When mounting the wire, avoid using ordinary staples because they might break through the insulation and short the wires inside. Insulated staples are available at most hardware stores. Tap them in gently with a small hammer. Avoid routing wires along heat ducts

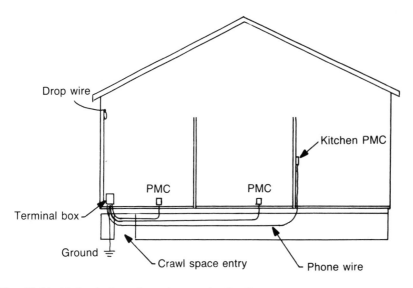

Fig. 10-11. Method of running wires under the floor.

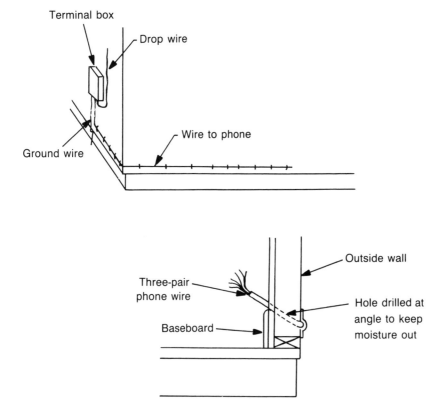

Fig. 10-12. Wire to telephone wrapped around outside of house to point of entry.

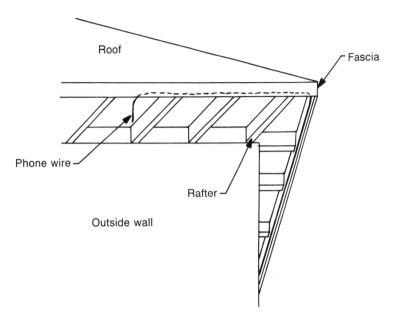

Fig. 10-13. Phone wires running underneath roof overhang.

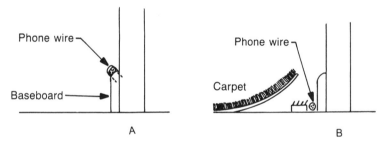

Fig. 10-14. Telephone wire can be run along the top of baseboards (A) or under the edge of the carpet (B).

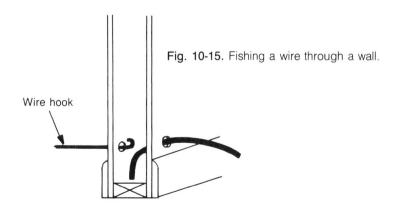

Fig. 10-15. Fishing a wire through a wall.

or metal pipes. In a few cases, metal objects might cause interference with sound quality.

It is good practice to complete the installation of the new telephone jacks and leave the connections to the power source for last so you are not working on a live circuit. Normally the voltage isn't enough to harm a healthy person, but if the telephone should happen to ring, the voltage could increase to between 75 and 100 volts. Simply take the other telephone off the hook. In addition, you might want to stand on a dry piece of carpet or some other insulating platform.

When connecting the wires to the telephone jacks, look closely at the jack. You will see letters indicating what color wire is attached to each terminal (Fig. 10-16). Red, green, yellow, and black are the standard colors. Modular connectors have either four or six copper pins. Four-pin connectors have pin numbers 2, 3, 4, and 5. Six-pin connectors have pin numbers running 1, 2, 3, 4, 5, and 6. In both cases, the two innermost pins are numbers 3 and 4. Connect these

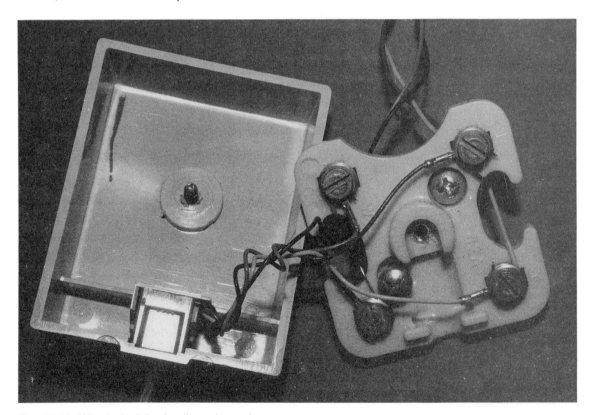

Fig. 10-16. Wire jacks following the color code.

two pins to the two terminals where the incoming pair are connected. If red and green wires are used at the terminal box, then the red and green wires at the jack should go to pins 3 and 4 of the modular connector. If you have six wires in your installation, you will have additional colors with combinations of blue, white, and orange. The important point is to stay with a color code.

If you have connected the right pair, you should be able to plug in the telephone, hear a dial tone, and call out. If you have a dial tone but can't call out, try reversing the wires at either the terminal box or the jack—one or the other but not both. Some telephones operate with polarity (a negative and positive) and the wires might need to be switched.

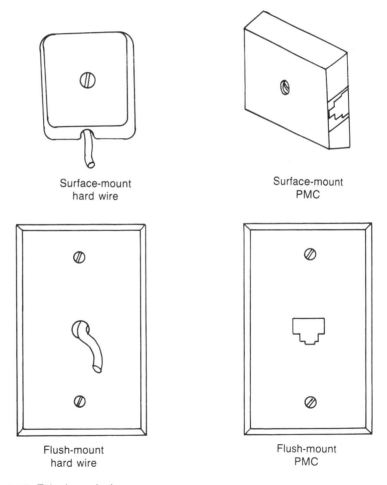

Surface-mount
hard wire

Surface-mount
PMC

Flush-mount
hard wire

Flush-mount
PMC

Fig. 10-17. Telephone jacks.

You can use pins 2 and 5 for a lighted dial or touch pad. Use the remaining black and yellow wires to connect to these pins. If your system uses six wires, you can use the last two wires for a separate telephone number or simply tuck them aside to be used as spares.

Wall jacks come in a variety of sizes and configurations. They can be surface or flush mounted (Fig. 10-17). A duplex jack is available for plugging in two telephones at the same location. There is even a weatherproof jack for an outdoors location by the pool or patio.

An installation kit is available from AT&T telephone centers for about $25. This kit is called Add-An-Outlet, and it contains 50 feet of wire and all of the necessary hardware to install three telephone jacks. The kit comes complete with instructions and a special wire stripper.

11

Troubleshooting Techniques

Usually the first sign of a problem is when something doesn't work—the microwave won't start or the television won't come on. A check at the circuit breaker or fuse panel often reveals a tripped breaker or blown fuse. One good thing about fuses is that the little window gives you a clue as to what happened. If the window is scorched, or darkened from heat, the failure is probably caused by a short circuit. The little metal strip inside is instantly heated to such a high temperature that it vaporizes and discolors the window. If the window is clear, the metal strip simply overheated and melted. The overheating is probably caused by an overloaded circuit. With circuit breakers, you only know something has happened.

There are other reasons circuits fail, such as loose connections and improperly seated fuses, but the most common problems occur from short circuits and overloads. A *short circuit* means that a shorter path to ground was provided instead of the desired circuit. Faulty appliance cords are a common source of trouble. Frayed or worn insulation allows a bare wire to touch another bare wire or grounded metal housing, producing a short circuit. The current suddenly increases beyond the capacity of the fuse or breaker, and the circuit fails.

You can usually find short circuits in a matter of minutes because most problems occur in flexible cords, plugs, or appliances. Look for blackened areas on outlets or charred or frayed appliance cords plugged into the dead circuit. Replace the damaged plug or cord, and replace the fuse or reset the breaker.

LOCATING THE PROBLEM CIRCUIT

If there is no visible evidence of any problem, unplug all the appliances, and turn off all wall switches on the dead circuit. Then replace the blown fuse or reset the breaker. If the new fuse blows or the breaker trips, remove the fuse or make sure the breaker is off. Now remove the cover and pull out the switches or receptacles one at a time. Look for charred insulation or burn marks caused by a bare wire touching the back of a metal box. The switch or receptacle itself might even be broken. Replace the damaged wire or broken switch or receptacle, and restore the power.

If the new fuse doesn't blow or the breaker trip, turn on the wall switches one at a time until the fuse blows or the breaker trips. The problem is probably in the fixture controlled by the switch or the switch itself. Make sure the circuit is dead, then inspect the fixture and switch. Look for the obvious black marks, charred wires, or burned wire connections. Repair bad connections or replace the faulty fixture or switch. Then restore the power.

If the fuse doesn't blow or the breaker trip and all of the switches are on, then the problem is in one of the appliances you unplugged earlier. Plug in and turn on one appliance at a time until the circuit fails. If the fuse blows or breaker trips when the appliance is plugged in, inspect the plug and cord, but if the circuit fails only when the appliance is turned on, the switch or the appliance itself is defective.

COMMON PROBLEMS AND FAULTS

It doesn't happen often, but sometimes the circuit wiring is defective. A staple might have been driven in too far and broken the insulation. The problem might not be discovered until condensation creates enough moisture to cause the short. To determine if the wiring is defective, you must remove all of the devices from the circuit. If the problem is still there, then the breaker might be bad or the wiring faulty. Try a new breaker first, but if that doesn't work you will need to run a new cable and abandon the old one. With the power off, use an ohmmeter or some type of continuity tester to determine which part of the circuit is shorted, then just replace that run.

By far, most short circuits are caused by faulty plugs and cords. The constant flexing and sometimes abuse causes them to wear quickly. Some of the danger signs relating to cords and plugs include burn marks around the plug, intermittent operation of the appliance, and physical damage of the cord when it is badly frayed or the insulation is brittle. In these instances, replace the old cord; don't repair it (Fig. 11-1). Excessive heat, when the cord or plug is warm to touch, is an indication of impending problems.

Power cords vary in size, but in order for them to be flexible, they all contain only stranded wire. Most plugs for small appliances have two prongs, but some will have three (Fig. 11-2A). The third prong will have a different shape; this is the ground connection. Another plug that may need attention is the *appliance plug* (Fig. 11-2B). It is surprising the number of faulty plugs that go unattended, when it only takes a few minutes to replace them.

Lamp sockets also can cause problems, but they are also easy to replace (Fig. 11-3). You can find a broken plug or loose prong through a simple inspection.

Take a look at the plug and receptacle. Is one prong larger than the other? (See Fig. 11-4) If so, the smaller prong or slot is the hot or live one. Make sure the black wire is connected to that side. Keep in mind that any connection that is not electrically and mechanically tight will generate heat and eventually cause a problem.

Receptacles are easy to replace. Buy a receptacle identical to the one you are replacing. If the old receptacle has only two slots, it is ungrounded and you should replace it with an ungrounded one. If the old receptacle has the two slots plus one round hole, replace it with the grounded type receptacle.

First make sure the circuit is dead, then remove the cover plate. Next remove the two mounting screws—one at the top, and one at the bottom. Now carefully pull the receptacle forward from its box. Pull it forward just enough to get at the connections. This way the wire is easier to fold back into the box later. If part of the receptacle is controlled by a wall switch, the break-off tab between

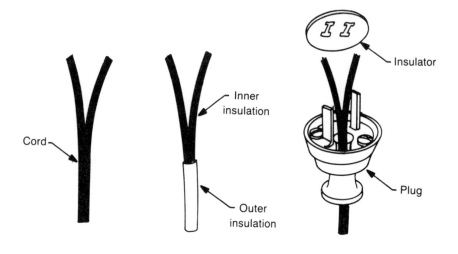

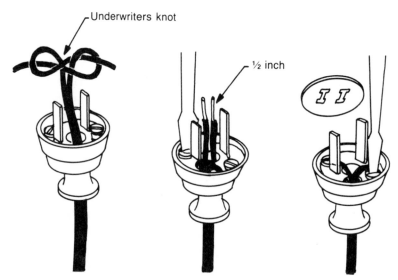

Fig. 11-1. To install a new plug, separate the cord about 2 inches. If there is an outer insulation, it will have to be removed first. Next slip the cord through the plug and tie an Underwriter's knot in the cord. Strip off about ½ inch of the insulation and loosen, don't remove, the terminal screws. Twist each of the stripped ends clockwise, then wrap each wire, also clockwise, around the terminal screws and tighten.

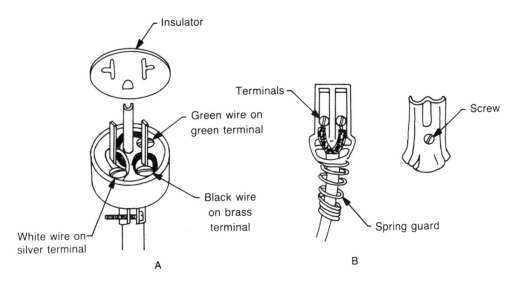

Fig. 11-2. A typical 3-prong plug (A) and an appliance plug (B).

the brass screws has been removed. You need to break the tab on the new receptacle to make it identical to the one you are replacing (Fig. 11-4).

If the wires are long enough, cut them from the old receptacle, strip the ends, and form new loops. Opening and closing the loops makes them brittle, so it is always better to work with new loops, if possible. Remove one wire at a time, and bend it to the side to help make sure they go back the same way they came off. The black wires go to the hot side, with the brass or dark screws, and the white wires go to the neutral side, with the silver colored screws. Most new receptacles have their screws already backed out, so loop the wire clockwise around the screw and tighten it. Don't use excessive force; very snug is what you want.

A number of today's receptacles offer the option of back-wiring (Fig. 11-5). In this case, the bare wire ends are straight. Simply insert them into the connection holes in the back of the receptacle. The hole has a clamping action, which secures the wire. A slot next to it is used to release the wire. Insert a small screwdriver in this slot to release the wire.

To replace a switch, first make sure the circuit is dead. Then remove the cover plate and the switch's mounting screws. Next,

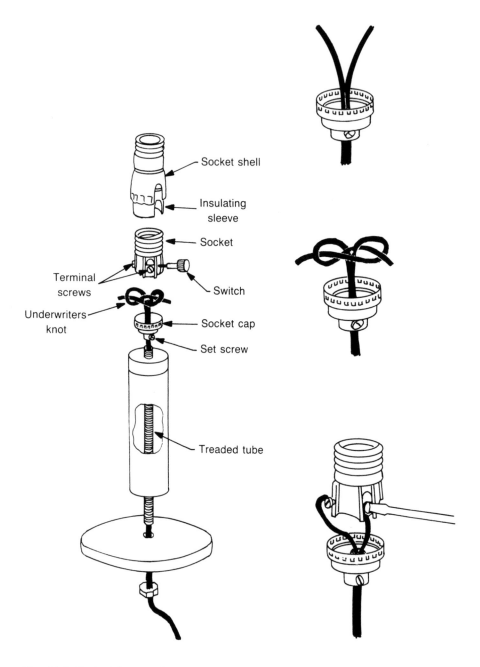

Fig. 11-3. To install a new lamp socket, slip the cord through the socket cap, tie the Underwriter's knot, strip ½ inch of insulation from the ends of the wire, and twist each end clockwise. Next loosen (don't remove) the terminal screws. Then wrap each wire clockwise around the screws and tighten.

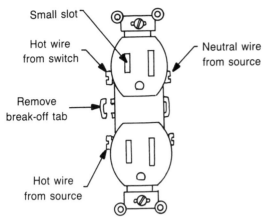

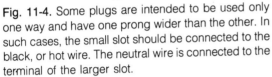

Small slot

Hot wire
from switch

Neutral wire
from source

Remove
break-off tab

Hot wire
from source

Fig. 11-4. Some plugs are intended to be used only one way and have one prong wider than the other. In such cases, the small slot should be connected to the black, or hot wire. The neutral wire is connected to the terminal of the larger slot.

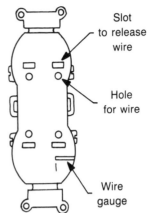

Slot
to release
wire

Hole
for wire

Wire
gauge

Fig. 11-5. The back view of a receptacle showing the holes used for back wiring.

carefully pull the switch forward from its box, just far enough to disconnect the wires. If the wires are long enough, cut them from the switch and make new loops. You might find a black wire and a white wire instead of two black wires. This may seem like a violation of the code, but a switch loop is the only case in the code in which the white wire can be used as a hot wire. Align the new switch so that when the toggle is up, the switch will be on, and then connect the wires. You might want to try the switch before you continue the installation. Turn the switch to the ON position, make sure it's clear, and restore the power. If the light works, turn the power back off, install the switch in the box, and replace the cover.

OVERLOADS

The other common cause of circuit failures is a circuit overload. This problem can sometimes be identified when the fuse blows; the little window remains clear. As you add appliances to a circuit the current increases to meet the demand. If enough appliances are added, the current will exceed the safe capacity of the circuit and the fuse will blow or the breaker trip. The kitchen is the most common area for this to happen. A microwave, a toaster, and an electric skillet on the same circuit is usually all it takes for this problem to occur. To solve the problem, the easiest way is to plug one of the appliances into an outlet from another circuit. The code requires that at least two 20-amp appliance circuits be installed to supply the kitchen.

If there are enough outlets, you might want to split the circuit and make two circuits (Fig. 11-6). The best way is to divide the circuit in half, if possible. You will have to run a new cable from the service-entrance panel. Pick out a suitable receptacle to separate the two halves of the circuit. Turn off the power to the circuit, and remove the receptacle from its box. You should find two white and two black wires, plus the ground wires, in the box. They will be the cable coming from the service-entrance panel and the cable continuing on to feed the outlets downstream. You'll have to guess which one is coming from the panel.

Disconnect one cable from the receptacle, separate the wires, and bend them back out of the way. You now have only one set of wires connected to the receptacle: one white, and the other black. Plug a lamp into the receptacle, and make sure that all of the free

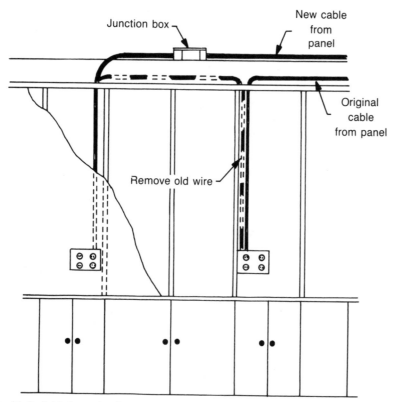

Fig. 11-6. A kitchen circuit with the receptacles split.

wires are well separated. Now turn the power back on. If the lamp lights, the cable from the panel is still connected to the receptacle. This is what you wanted. If the lamp doesn't light, then turn the power back off, reconnect the wires you disconnected, and disconnect the other wires. Make sure the wires are clear, and see if the lamp lights this time. It should. If the receptacle worked before, one of the two cables coming into the box must feed the receptacle.

After determining which cable is which, turn off the power, loosen any cable clamps, and pull the cable feeding the downstream receptacles from the box. This process will be easier if you have a helper. You might need to pull the cable from the attic or the basement. Have someone jiggle the selected cable so you will know which one to pull. Do not jerk the wire. Instead, apply steady pulling pressure.

Once you have the cable feeding the second half of the circuit free, run the new cable from the service-entrance panel to a point convenient to joining the two cables. If the old cable was damaged when it was removed, cut the damaged part off. Securely mount a junction box and use wire nuts to connect black wires to black and white wires to white. Connect the ground wires together. If the box is metal, also connect a bare pigtail ground wire to the box.

Make sure each wire is securely connected, then install the cover. You can put the receptacle that was left loose back in its box and install the cover after you retighten any loosened cable clamps. Connect the new cable to an appropriately sized breaker in the service-entrance panel and restore the power.

Appendices

Appendix

Material Costs

After you have decided on a project and that you can do the job yourself, probably the next question will be how much will it cost. The finished cost of any project is reflected mostly in labor with a small portion going to materials. Wiring projects are a good example. Because of more efficient manufacturing due to competition, and the increasing use of plastic boxes instead of metal, electrical hardware has remained inexpensive. The price of copper wire fluctuates with the world price of copper, but the copper wire used in residential wiring (12-2 and 10-2) tends to be a leader item for the electrical supply houses and remains relatively stable. To a large electrical contractor competing in the market and using hundreds of feet of wire a day, the cost of material is watched closely. For the homeowner, however, the wiring job will be smaller and materials will be fewer. A material increase of 10 percent to a contractor could mean a cost difference of hundreds of dollars, while a $50.00 material bill to a homeowner would only increase $5.00. This should not be enough to discourage anyone wanting to improve their home electrical system.

To estimate a job, count the outlets, add the cost of the fixtures, then estimate the amount of wire needed. Add about two feet extra for each outlet for mistakes and to make up the connections.

For example, to install two three-way switches for an existing hall light, you might expect to spend about $15.00 for two switches. Twenty-five feet of 12-3 wire will cost about 33 cents a foot, bringing the wire cost to about $8.25. This brings the cost to about $23.25. Using the existing switch box, you'll need one more switch box and plate—about another dollar. this brings the total to about $25 for material. You might get an electrician to do the job in his off time for $100, labor and material.

Labor costs vary considerably depending on the area you live, but material cost should be about the same.

The following price list is for a general reference only. The values listed are approximate. Contact local electrical supply houses for a more precise figure.

MATERIAL COSTS

Tools

Cable ripper	$1.50
Channellocks	$20.00
Crimpers	$10.00
Dykes	$12.00
Lineman's pliers	$13.00
Needlenose pliers	$5.00 to $8.00
Neon tester	$2.00 to $5.00
Screwdriver	$3.50 to $4.50
Stripper/crimpers	$4.00 to $8.00
Tape measure	$8.00 to $10.00
Volt-ohmmeter	$15.00 to $85.00
Wiggy	$20.00
Wire stripper	$3.00

Service-Entrance Panels with Main Breaker Installed

100-amp indoor w/o circuit breaker	$100.00
200-amp rainproof	$140.00
200-amp indoor	$220.00
200-amp rainproof	$260.00
Subpanel for shop	$20.00 to $30.00

Circuit Breakers

	One Pole	Two Pole
15 amp	$6.00 to $8.00	$22.00
20 amp	$6.00 to $8.00	$22.00
30 amp	$6.00 to $8.00	$22.00

GFCI Circuit breaker	$55.00 to $65.00 (20 amp), depending on the brand
Weatherproof head	$3.00 to $6.00 (plastic)
LB condulet	$5.00 to $6.00

Boxes (Plastic)

Switch/receptacle	$ .30
Two-gang switch/receptacle	$1.30
Light fixtures	$1.30

Boxes (Metal)

Switch/receptacle	$ .84
Two-gang switch/receptacle	$2.25
Light fixtures and junction	$1.50 to $1.75

Receptacles

Duplex	$3.50
Air condition	$8.00
Range	$10.00
GFCI	$15.00 to $55.00

Switches

Single pole	$2.50 to $4.00 (quiet type)
Three way	$7.00 to $8.00
Four way	$14.00
Dimmer switch	$8.00 to $12.00

Wire

Size	Price per Foot	Price per Roll
14-2 WG NM	$ .15	$19.00 to $22.00 (250 ft.)
12-2 WG NM	$ .19	$25.00 to $30.00 (250 ft.)
12-2 WG UF		$10.00 (50 ft.)
12-3 WG NM	$ .33	$77.00 (250 ft.)
12-2 WG NM	$ .41	$77.00 (250 ft.)
10-3 WG NM	$ .59	$100.00 (250 ft.)
8-3 WG NM	$1.33	$124.00 (125 ft.)

Miscellaneous

Wire nuts	$7.00 for a box of 100
Electrical tape	$1.50 per roll
Telephone wire	
2 pair (4 wires)	$10.00 (100 ft.)
3 pair (6 wires)	$14.00 (100 ft.)
Telephone jacks	
PMC	$3.00 to $4.00
Ceiling fans	$40.00 to $150.00
Door chimes	$20.00 to $70.00
Garage-door openers	$150.00 to $200.00
Low-voltage	
yard lights	$45.00
Smoke detectors	From about $10.00 for a battery type to $35.00 for a 120-volt. Add $10.00 to $15.00 more for the battery backup.
Thermostats	$25.00 to $35.00
Water-heater timers	
100 volt	$35.00
220 volt	$40.00

Glossary

Glossary

alternating current—Current that regularly reverses its direction, flowing first in one direction then the other. Abbreviated AC or ac.

ampacity—The amount of current, expressed in amps, an electrical conductor can carry continuously without exceeding its temperature rating.

ampere—A unit used in measuring electrical current, abbreviated **amp**. It is based on the number of electrons flowing past a given point in one second. It can be determined by the Ohm's Law formula, I (current) = watts divided by volts.

AWG(American Wire Gauge)—The adopted standard of wire sizes such as No. 12 wire, No. 14 wire, etc. The larger the number of the wire, the smaller the size of the wire. A No. 14 wire is smaller than No. 12 wire.

bonding—The permanent joining of metal components of an electrical system to form a continuous, conductive path capable of handling safely any current likely to flow.

bonding jumper—A reliable conductor between metal parts where the parts may be temporarily separated at some time.

bonding jumper, main—The connection between the neutral bar in the service entrance panel and the panel.

branch circuit—Any one of a number of separate circuits distributing electricity from an overcurrent protection device.

busbar—The solid metal conductor in a distribution panel that provides a common connection between the main service and the branch circuits.

cable—A stranded conductor or a group of individual conductors insulated from each other.

cable, entrance—A heavy cable used to supply electrical service from the main line to a building.

cable, nonmetallic sheathed—Two or more insulated wires assembled inside a plastic sheath. Type NM is used in dry locations. Type NMC may be used in both dry or damp locations.

circuit—An electrical conductor forming a continuous path, allowing current to flow from a power source through some device using electricity and back to the source.

circuit breaker—A safety switch installed in a circuit that automatically interrupts the flow of electricity if the current exceeds a predetermined amount. Once tripped, a circuit breaker can be manually reset.

conductor—The trade name for an electric wire or busbar capable of carrying electricity.

continuity—The state of having a continuous electrical path.

current—The transfer of electrical energy caused by electrons traveling along a conductor, abbreviated I.

cycle—One complete reversal of alternating current where a forward flow (positive) is followed by a backward flow (negative). The standard rate in the United States is 60 cycles per second, now called 60 Hertz.

duplex receptacle—A receptacle providing electrical connections for two plugs.

fuse—A safety device installed in a circuit designed to interrupt the flow of electricity should the current exceed a predetermined amount. Fuses cannot be reused.

ground—A conducting connection between earth and an electrical circuit or equipment, whether intentional or accidental.

ground clamps—Metal clamping devices used to connect a wire to earth ground, often through a ground rod or water pipe.

grounding electrode conductor—The conductor from the neutral busbar in the service entrance panel to the established ground.

ground-fault circuit interrupter—A safety device installed in a circuit designed to protect personnel by detecting very small currents and interrupting the flow, abbreviated **GFCI**.

hot busbars—The solid metal bars in service entrance panels and subpanels where the power source is connected. Circuit breakers or fuses mounted on these busbars deliver power to individual branch circuits.

hot wire—The ungrounded current-carrying conductor of an electrical circuit. It is normally identified by black or red insulation, but it can be any color except white, gray, or green.

insulation—A nonconducting material used to cover wires and components to remove shock hazards and to prevent short circuits.

junction box—A box containing only the connections or splices of several wires.

kilowatt—A unit of electrical power measured as 1000 watts, abbreviated **kw.**

kilowatt-hour—The unit used in metering electricity. One kilowatt-hour is 1000 watts used in one hour or the equivalent, such as 500 watts used for two hours. Abbreviated **kwh.**

knockout—Round, partially punched-out opening installed by the manufacturer in panels and junction boxes that allows the opening to be knocked out with a screwdriver.

low-voltage wiring—A method of wiring where a lower voltage is used to supply electricity for such purposes as doorbells, thermostats, and some outdoor lighting.

neutral busbar—The solid metal bar in a service entrance panel or subpanel used as a common terminal to connect all of the neutral wires. The neutral busbar in the service entrance panel is bonded to the panel as well as being directly connected to earth ground. The neutral busbar in the subpanel is used to connect the neutral wires, but it is isolated from the panel and is not grounded.

neutral wires—The grounded conductor that provides the return path to the source, completing the circuit. Neutral wires must never be interrupted by circuit breakers or fuses. Neutral wires are identified by white or gray insulation.

ohm—The unit for measuring electrical resistance.

ohmmeter—The instrument used in measuring electrical resistance.

outlet—A point in a wiring system where current may be taken to supply electrical equipment.

overcurrent protection device—A fuse or circuit breaker that automatically interrupts the flow of electricity in the event the current exceeds a predetermined amount.

overload—A situation where an electrical circuit is attempting to carry more current than it can safely handle.

pigtail—A single wire extending from a connection of two or more wires.

receptacle—A connecting device designed to accept plugs.

resistance—The property in an electrical conductor or circuit that restricts the flow of current. It is measured in units of ohms.

service drop—The overhead wires from the utility pole that deliver the electricity to the building.

service entrance panel—The main power cabinet containing the main breaker and circuit breakers distributing electricity throughout the residence.

service lateral—Underground wires providing electrical service to the building.

short circuit—A completed, very low-resistance circuit where two bare hot wires come in contact or a bare hot wire touches a bare ground wire or grounded component.

source—The point supplying the electrical power. It may be a battery, generator, or the service entrance panel.

switch—A device used to close or open a circuit, allowing current to either flow or not flow, respectively.

terminal—A point used to make electrical connections.

thermostat—A device used to control temperatures to a predetermined level.

transformer—A device used to transfer electrical energy from one circuit to another using electromagnetic induction. The two types commonly found are the step-up and step-down transformers. The step-up transformer is used where a higher output voltage than the input voltage is desired. A step-down transformer is used for circuits where a lower output voltage than the input voltage is needed, such as with doorbells and thermostats.

upstream/downstream—The location of a point in a circuit relative to the power source. Upstream means the part of the circuit between the point and the power source. Downstream refers to the part of the circuit from the point to the remaining part of the circuit going away from the power source.

volt—The unit used in measuring electrical pressure. Abbreviated **V** or **E**.

voltage—The electrical pressure, measured in volts, at which a circuit operates. Voltage can be determined by the Ohm's Law formula, watts divided by current = voltage.

voltage drop—An electrical term describing the loss of voltage that occurs when the wires are not of sufficient size to carry the amount of current flowing.

voltmeter—A meter used to measure voltage.

watt—The unit used in measuring electrical power. Abbreviated **W**. The amount of power in a circuit can be determined by the Ohm's Law formula: w = current × voltage.

Index

Index

The Book Club offers a wood identification kit that includes 30 samples of cabinet woods. For details on ordering, please write: Grolier Book Club, Inc., Member Services, P.O. Box 1785, Danbury, CT 06816.